CHEMICAL AND PROCESS ENGINEERING SERIES

Series editor: I. L. Hepner

Odour Pollution of Air
Causes and Control

W. SUMMER F.Inst.E., M.Inst.P.,
M.Amer.Phys.Soc., Hon.F.Phys.A., M.R.S.H.

LEONARD HILL
LONDON 1971

An Intertext Publisher

Published by Leonard Hill Books
a division of
International Textbook Company Limited
158 Buckingham Palace Road
London SW1

First published 1971

ISBN 0 249 44022 9

Printed in Great Britain by
Billing and Sons Ltd, Guildford and London

Odour Pollution of Air
Causes and Control

Preface

In all industrialized countries, the problems of the polluted environment
have become acute. Legislation now provides, in most of them, a legal
basis for complaints and the rectification of nuisances, for compensation
for losses and for offence caused. With a new awareness, the citizen has
begun to react against soot and smoke and noise and smell, all things that
have been until quite recently accepted as a more or less inevitable
background of town life. The subject has become a major concern of
governments, administrators, chemists, engineers and town planners. It
concerns all the construction industries; it has for generations been a
matter for lawyers; and in Europe certain states have even been guilty of
poisoning each others' air and water.

Pollution by smell, the subject of this book, is admittedly the least
directly harmful to health of all the environmental nuisances. But it can
seriously affect mental attitudes and destroy the peaceful enjoyment of
our homes and neighbourhoods, while it is often the sign of a wasteful
process. At all events, a process of the kind society will no longer tolerate.

An intriguing aspect of this study is its partly subjective character. For
while we have instruments that will measure with accuracy the density of
smoke, the size of grit particles, the intensity of noise or the composition
of engine exhausts, only the human nose can 'measure', or even say what
we mean by, smell. There is no instrument available to make man-related
measurements of odour, no possibility of expressing the quality of an
aroma in such terms that everybody would recognise it forthwith. Its
quantity may be approximately described in terms of chemical concentration, but even this must not be regarded as having any bearing on its
perception.

The amelioration of the present conditions is an urgent necessity. Signs
are not wanting that improvements in production, and of concept, are
arriving.

This book aims at supplying both theoretical and practical background for coping with all the main problems of smell, its production, sensation, and abatement.

It is not possible without intolerable repetition to discuss in a book of this kind technical details of every process or branch of industry which is producing, or plagued by, offensive odours. It has been found expedient first to discuss general principles, then, in the practical sections of the book, some typical applications of each principle. The engineer will be able to select the correct method, or technique, in accordance with the characteristics of each case.

W. SUMMER
September, 1971

Acknowledgements

It is my pleasant duty to acknowledge the interest Mr P. Edmonds has taken in the editing of the book. His advice proved most valuable, and it is with sincere thanks that I record this fact.

I would like also to place on record my grateful indebtedness to the following who have gone out of their way to provide information on specialized items:

Air Pollution Control Association, Pittsburgh, Pa.
Barnebey Cheney, Columbus, Ohio
Brightside Ltd., London
Brookhaven National Laboratory, Meteorology Group, Upton, N.Y.
Central Electricity Generating Board, London
CJB Processes Ltd., London
Brødr. Hetland, Bryne, Norway
Farbwerke Hoechst A.G., Frankfurt/Main
W. C. Holmes & Co., Ltd., London
Japan Air Cleaning Association, Tokyo
Manufacturing Chemists' Association, Washington
A. S. Myrens Verksted, Oslo
Pennsylvania State University, Pennsylvania
Sewerage Board, States of Jersey, C.I.
Siemens A.G., Erlangen
Société de la Revue *La Pollution Atmosphérique*, Paris
Vickers Ltd., Engineering Group, London
Warren Spring Laboratory (Ministry of Technology), Stevenage, Herts.

To my Wife

Contents

Preface   v
Acknowledgements   vii
List of figures   xi
List of tables   xv
Conversion factors   xvii
Introduction   xix

Part I   FUNDAMENTALS
 1   A world problem   1
 2   Some hypotheses of odour   9
 3   Thresholds of olfaction   18
 4   Aspects of odour   30
 5   Pathogenic effects of odour pollution   38
 6   Methods of detection and measurement   47
 7   Social aspects of odour   57
 8   Pure air and stale air   59
 9   Weather   72
10   Vegetation and air pollution   73

Part II   LEGAL ASPECTS AND EVALUATION
11   Legal aspects   76
12   Panels and evaluation   86
13   Odour evaluation in court cases   97
14   Anomalous odour perception   100

Part III   INDUSTRIAL SOURCES OF ODOUR POLLUTION
15   Meat and fish processing   104
16   Tanneries   109
17   Tallow melting and refining   111
18   Gut cleaners   113

19   Tripe boilers   115
20   Processing of white fish   117
21   Cooking smells   121
22   Catering industry   126
23   Fish friers   128
24   Merchant shipping   129
25   Maggot breeding   134
26   Sewage and domestic refuse   137
27   Pharmaceutical industry   151
28   Oil refineries   152
29   Paper mills   156
30   Other industries   158

Part IV   TECHNIQUES OF AIR DEODORIZATION
31   Phase separation   162
32   Wet-scrubbing   164
33   Dry-scrubbing   178
34   Chemical cleaning   199
35   Dilution   205
36   Electrostatic precipitation   230
37   Condensation   232
38   Masking   235
39   Combustion   239
40   Ultra-violet irradiation   244
41   Deodorization by change of process   261

Part V   PREVENTION OF ODOUR POLLUTION
42   Town Planning   272
43   The odourless factory   275
44   Air pollution prevention service   281

References   287
Appendix: Organizations   297
Index   303

Figures

1 Relationship between rise in world population and world industrial productivity, expressed as consumption of electric power   5
2 Block diagram representing the principle of olfaction   9
3 An osmoceptor   12
4 Graphical representation of Weber's Law, showing the odour threshold or latent period   26
5 The basic scheme of the olfactory system   30
6 The effect of boiling point on volatility   32
7 (a) Radiographic appearance of complicated pneumoconiosis in a 39-year old man who had worked in the graphite industry for 20 years (b) Graphite bodies lying free in an alveolar space   39
8 Extraction unit which purifies the air from a 'Chayen process' bone fat extraction plant   106
9 Deodorizing equipment for direct rotary driers   119
10 Deodorizing system for steam driers   119
11 An arrangement of ultra-violet tubes for the destructive deodorization of air working in conjunction with an air conditioner in a restaurant kitchen which has no outlet for the polluted air   125
12 The changing density of refuse in a typical British city   137
13 The compost house of Bangkok's refuse composting plant   138
14 Flow diagram of the Jersey, C.I., refuse treatment plant   140
15 Loading cell, Jersey, C.I.   141
16 Fermentation building, Jersey, C.I.   141
17 A Vickers seerdrum   142
18 Wet-scrubbing: (a) Polluted air and water spray move in a countercurrent (b) Polluted air and water moving concurrentwise   165

19   The Kinpactor, a wet-scrubber using the venturi principle   173
20   Drawing of a type 'JC' multi-wash collector, and sectional view   175
21   Sectional view of the Holmes-Schneible type 'SW' cupola collector   176
22   Wet-scrubber without nozzles   176
23   The size of particulate matter which a filter will pass is a function of the optimum radius for penetration of a particle and of the mean air velocity in the fibrous filter   180
24   Snifter set   186
25   Installed plant cost for solvent recovery   187
26   A plant for the absorption, on activated carbon, of an organic solvent (136)   193
27   Useful life of activated carbon   195
28   A drier using the helical flow technique. Helical air currents carry the dust along to the dust collector   196
29   A drier using the helical flow technique. Annular stationary air currents provide the required residence time   197
30   Celmar extraction system installed at Courtaulds Ltd., Preston   198
31   Stack exit velocity v and wind velocity c shape the plume emanating from the chimney   207
32   Stack exit velocity v and wind velocity c determine the rise angle of the plume   208
33   Effect of atmospheric instability on diffusion from a chimney   210
34   Wind tunnel test on multi-flue stack : (a) Downwash occurs with flush-topped design (b) Improvement in emission with projecting flues   211
35   The lower strata of a plume have the tendency to cause an odour nuisance even at a considerable distance from the stack. The upper strata are taken up by the winds   212
36   Winds fanning out over a ridge or hill and across a valley come down into the valley after hitting the opposite side of the hill   213
37   Diagrammatic representation of plume dispersion under different conditions of atmospheric stability   215
38   Plumes from the 320 ft (nearly 100 m) high stacks at Deptford Power Station, London, dispersing above the fog surface, 750/800 ft (about 250 m) above ground level   216
39   Houses of a newly developed hillside estate were so distributed in terraces that foul air from the septic tanks of one row of houses blew straight into doors and windows of the houses above   217

40 Uncorrected chimney heights:
    (a) Very small installations   219
    (b) Small installations   220
    (c) Medium installations   221
    (d) Large installations   222
    (e) Final chimney heights   223
41 The geometry of a point source   249
42 The geometry of a linear source   250
43 Irradiation of a vertical wall by a linear source   251
44 Two identical rooms are illuminated with identical sources
    except for the fact that in (a) the rays are not controlled in their
    distribution whereas in (b) the sources are situated in reflectors
    which restrict spread of radiation   252
45 Arrangement of sources of ultra-violet radiation in recesses
    across duct or irradiation chamber, if particulate matter as
    well as offensive fumes are carried   253
46 Effect of distance of a source from the ceiling on volume of
    irradiated air, and on amount of nascent oxygen and ozone
    produced   257
47 Effect of differently shaped reflectors on the distribution of
    radiation   257
48 Some possible flow sheets   262
49 Flow sheet of a CS_2–H_2S recovery plant, system Pintsch-Bamag 268
50 Change in wind direction with height   273
51 General scheme for the treatment of polluted air   277
52 General plan of the air pollution control system at Farbwerke
    Hoechst A.G., Frankfurt/Main—Hoechst   283
53 (a) Instrument van with experimental plants on the roof
    (b) Instrument van, interior view   283
54 (a) TV camera for the observation of the air space above the
    works   284
54 (b) TV screen and control box   285

Tables

1	Factors of population increase for the year 2000	2
2	Major regional population forecasts 1960–2000	3
3	World total population forecast	3
4	Historical trend of world population	4
5	Specific consumption of energy	6
6	Average national income per capita	6
7	Social (industrial and economic) potential	7
8	Characteristic specific frequencies of chemical linkages	14
9	Olfactory thresholds and maximum acceptable concentrations of some air pollutants	22–4
10	Osmogenic and nasal irritation threshold	28
11	Vapour pressure of osmogenes	33
12	Sensory functions of cranial nerves	35
13	Size distribution of pulmonary dust	41
14	Some vapours whose MAC-values are lower than their olfactory threshold	43
15	Composition of pure, dry atmospheric air	60
16	International standards for drinking water	62
17	Composition of respiratory air	63
18	Emission of thermal energy from the human body	68
19	Heat losses of the human body	68
20	Occupation and heat transfer	69
21	Effect of heat loss	70
22	Heat absorption of polluted air	71
23	Formation of methaemoglobin by ultra-violet radiation	131
24	Average composition of sludge gas	143
25	Biological effects of hydrogen sulphide	145
26	Effective biochemical oxygen demand and temperature	146
26a	Minimum velocity and effective biochemical oxygen demand	147

27 Industrial solvent odours   159
28 Average BOD values of industrial wastes   166
29 Aqueous solubility of some osmogenic substances   167
30 Solubility and temperature   167
31 Analysis of sea-water   170
32 Pressure exerted on air filters   180
33 Pollutant concentration and filter material   181
34 Penetrability of filters   181
35 Sorptive power of activated carbon at NTP   190
36 Surface-to-mass ratio of sorbents   191
37 Retentivity of activated carbon   192
38 Height coefficients   224
39 Beaufort scale of wind velocities   224
40 Wind speeds at various heights   225
41 Recommended air exhaustion data   227
42 Specific gravity of some osmogenic gases and vapours   228
43 Requirements of cooling water   233
44 Effect of cooling by condensing   233
45 Combustion properties   239
46 Flammable concentrations of gases and vapours in air   243
47 Spectral emission from a low-pressure generator   245
48 Intensity and distance of straight line sources   248
49 Relative conjunctivitis factor   256
50 Anti-corrosive protective coatings   260

Conversion factors

To convert	into	multiply by
British thermal unit	gram-calorie	251·996
Centimetre	inch	0·393
	foot	0·033
cm/sec	ft/sec	0·033
cmHg	inch H_2O	5·352
	lb/sq. inch	0·193
	kg/m^2	135·951
Cubic cm	cubic inch	0·061
	litre	0·001
Cubic ft	m^3	0·028
	litre	28·316
Cubic inch	cm^3	16·387
	litre	0·016
Cubic metre	cubic inch	61 023·375
	cubic foot	35·314
Ft/min	m/sec	0·005
	km/h	0·018
Ft^3/min	m^3/h	1·701
Inch Hg	atm	0·033
	kg/m^2	345·316
Inch H_2O	cmHg	0·187
	kg/m^2	25·400
Miles/hour	m/sec	0·447
kcal	kJ	4·184

Introduction

Air pollution is not a feature of the industrial age, although it is often linked with the beginning of the 'industrial revolution', i.e. the second half of the 18th century. The emission of soot and smoke from domestic chimneys was a nuisance to Londoners even in medieval times. Richard III (1377–99) taxed heavily the use of coal, and Henry V (1413–22) instituted a stringent control over its movement into London. Two centuries later, in 1661, John Evelyn addressed a dissertation called *Fumifugium: Or the Inconvenience of the Aer and Smoake of London* to Charles II. In this pamphlet, Evelyn suggested some remedies, which however did not reduce the nuisance (1). A person called Justel presented before the Philosophical Society 'An Account of an Engine that Consumes Smoke', 1686, but his invention was quite unsuccessful.

The buoyancy of particulate matter, whether solid or liquid, floating in air, is affected by many factors, not the least important of which is wind. The smallest particles may remain suspended in the atmosphere for very long periods of time and thus increase the density of pollution more in the higher strata of the atmosphere than in the lower. This may, in time, have serious climatic consequences, because the particles reflect some of the incident solar energy. Again, thermal equilibrium on our planet is maintained by the atmospheric carbon dioxide which is opaque to a certain range of wavelengths in the infra-red, but transmits luminous solar radiation. As its amount is slowly increasing, it would be a justifiable conclusion that the temperature of the lower atmospheric strata is in consequence slowly going up.

Yet recent temperature measurements indicate the contrary: world-wide cooling, not heating, is taking place. From investigations by workers at the National Centre for Air Pollution Control, Cincinnati (2), a very sorry picture develops; since the turn of this century the air over a populated place like Washington has been 'enriched' by about 28×10^6

aerosol particles per cm^2 of ground, assessed for the height of the atmosphere. Even in Davos, the Mecca for sufferers from pulmonary tuberculosis, the annual growth rate of aerosols is more than 10^7 per annum per cm^2. It is this 'dusty' atmosphere which causes the net cooling down of our environment.

Human activities produce, and release into the air, combustion products which largely increase the numbers of artificial condensation or 'Aitken' nuclei. These are generated naturally by the action of suitable radiation, e.g. light, on volatile matter released by vegetation. They are macromolecules which disappear again by agglomeration and precipitation; and the volatile matter consists, in the main, of terpenes (3).

We are not here concerned with the nuisance of air 'pollution' by noise. Ignoring such extremes as supersonic bangs from aircraft, noises of ordinary human occupation have greatly multiplied over the last fifty years.

The third great atmospheric offender is smell. In most cases it is a question of an obnoxious smell, one which causes discomfort and distress, but sometimes it is an odour which, under different conditions, might be considered pleasing.

Of these three scourges of civilization, solid and liquid particulate air pollution (aerosols and mists), and the pollution of air by noise, have been given legal formulations in Britain (*Clean Air Act*, in force since 5th July, 1958; and *Noise Abatement Act*, in force since 27th November, 1960) (4, 5). These are based on technically defined standards of pollution by smoke (Ringelmann Chart) (6) and by noise (sound level meter) (7). The definitions have been accepted by Parliament and have been included in the respective legislation.

There are no units or systems by which to express the pollution of air with osmogenic matter: the experts have failed to suggest, so far, an acceptable unit and method of measurement, or to construct an instrument for objectively measuring smells. Chemical trace-detection methods, such as chromatographic analysis, cannot replace organoleptic tests. The latter, however, are by no means objective, repeatable or reliable: they depend on numerous imponderabilia of human behaviour. The only time the law can interfere in the highly complex matter of human chemical production *v.* human comfort is by invoking the old definition of 'nuisance', and leaving it to the judge to decide whether or not such has been committed.

There are many other aspects of osmogenic air pollution which are, or will be, of importance to the individual and to society. As the population increases, the need for synthetic products will grow, at least in the same proportion, but probably faster. The national income per head is on the increase, and so is the variety of individual demand, which industry is eager to satisfy.

Much contemporary town planning shows vision and understanding. Regions for light, mixed and heavy industry are set apart, and the workpeople are housed close to their places of work. Yet the attempt to please everybody, the economist and the builder, the nature conservationist and the architect, does not always result in satisfaction in relation to the problems we are considering.

Industry—under the guidance of well-defined laws and at great expense —keeps the air free from smoke and soot, controls vibration and noise, and treats its harmful effluents. In the United States of America this will cost some \$275 000 000 000 between now and the year 2000. Of this enormous sum, two-fifths will be used in an attempt to halt, and reverse, water pollution; a similar amount will go in attempts to stop, and limit, air pollution; and the last fifth will serve the disposal of solid waste (8).

Yet, no provisions are made by manufacturers, or required by law, to *prevent* smells either from occurring or, if inevitable, from spreading to other areas. Local Authorities, having the power to penalize others, are yet often the worst offenders themselves with their sewage treatment works. Neither planners nor Authorities take into consideration the simple truth that 'wind knoweth no boundaries'. If and when a sufferer is successful in his claim for a reduction of his rates, £5 is the usual maximum concession. In each case the Local Authority agrees—without ever saying so—that the complainant has proved his case, that there is an offensive smell, that it is where it should not be, and that the Local Authority cannot, or will not, do anything about it.

It is not only the sewage works which give offence. Waste processors, garages burning old rubber tyres, chicken breeders trying to get rid of the bird manure either by burning it, or drying it in a kiln, sausage skin manufacturers, and the chemical and biochemical industry, all are potential, and very often actual, producers of obnoxious smells.

Against such pollution, at best, the condition will be written into an agreement or permit or licence that the trade must be carried on in a place so many miles away from town, and that it must not give offence to residents. Such a clause is inserted only in cases of officially confirmed and recognized offensive trades as specified in *Statutory Instrument* 1950, No. 1131 (9). Classes (iv) and (viii) are of special interest, the former comprising such trades as bone boilers, breeders of maggots from putrescible animal matter, chitterling boilers, fat melters, fellmongers, fish oil manufacturers, glue makers, and many more. Class (viii) is named in this *Statutory Instrument* the Special Industrial Group D amongst which are listed the petrochemical industry, the plastics, paint, and lacquer industry, the reclamation of rubber, and others. Group D is a summary of those industries which emit obnoxious smells and must be housed on estates well removed from residential, or even light industrial, areas. How, and how far removed, is left to the discretion and wisdom of the planners. The *Statutory Instrument* 1950, No. 1131 presents merely a classification of industrial occupations from the point of planning for human comfort, but offers neither advice nor refers to a Code of Practice as a guide to the abolition of smell nuisance caused, or likely to be caused, by these industries.

Administration requires only Classification. Town planners and engineers, however, want to know more about smells before they can offer designs of factories and industrial estates which, theoretically in Group D, no longer emit offensive odours.

To the ventilating engineer whose task it is to deal with a given odour problem, aesthetic aspects and differences of opinion carry no weight. To him, the problem usually resolves into two steps: first, to decide upon the method of odour control most appropriate to the case in point and, second, to overcome the financial argument that capital spent on "waste" is wasted. There are two fundamental approaches to odour control: (i) to avoid the production of smell, which may be possible, or partly possible, by altering that stage of the manufacturing process which causes the smell; and, should that not be feasible, (ii) to reduce the chances of the general air becoming polluted with air from the production areas, and to treat the latter by appropriate means before discharging it to open atmosphere.

It is regrettable practice to call in the ventilating engineer after the plans of a new factory have been completed. Considerable savings can be achieved if ventilation is introduced at the stage where the manufacturing processes are decided, flow sheets constructed, and machinery is sited. Production sections can be so laid out that the separation of foul from unpolluted air is achieved with a minimum of effort and cost; that processing plant giving off obnoxious odours is completely enclosed and connected to the ducted system carrying the foul air; and that odour control points are arranged in accordance with the requirements of the method used.

To expect to eliminate smells purely by redesigning manufacturing processes is a fruitless idea for reasons of economics except where, as in one or two cases, the odoriferous by-product is of reasonable commercial value. A sounder approach is to plan a new town, or new industrial estate, on the principle of the odourless factory, i.e. a factory in which generation, distribution and destruction of smells are as exactly controlled as are the manufacturing processes. Town and Country Planning will only be really successful when air pollution can be effectively controlled. Control means, in the case of particulate matter, prevention; in the case of noise, insulation; and in the case of smell, destruction. Feasibility has been demonstrated in all three groups. If planned from the beginning as an essential item of design, freedom from air pollution, especially by smells, is not an uneconomic factor, not a burden on production, nor a waste of capital.

Part I

FUNDAMENTALS

1 A world problem

Mankind, it is everywhere being ruefully observed, has shown little respect for the earth he inhabits. He has defaced its surface, has fouled its rivers, and polluted the air surrounding it. This melancholy picture is not made brighter or more promising by considering statistical forecasts for the next thirty years. They suggest that vehicular pollution alone, while discharging 'only' 70 million tons of various chemicals, many of which are carcinogenic, into the air around us, will increase by nearly one-half to 100 million tons in 1980 (10) and to twice this quantity by the turn of the millenium. Of course, devices to control air pollution from cars have been invented in great variety, and some countries have introduced legislation to make uncontrolled discharge from cars a punishable offence. Although by no means all cars in California—the first state in the United States to require antipollution devices on cars by law—are equipped yet, $2 \cdot 5 \times 10^6$ litre of unburnt petrol, and 2400 t of carbon monoxide are already prevented each day from being discharged into the air (11).

Taking the annual total of pollutants released into the air over the USA by motor vehicles of all descriptions, power plants, industry, domestic fires, and municipal as well as private refuse disposal, the National Bureau of Standards comes to the truly frightening figure of 140 000 000 tons (12).

In Britain, domestic fires and industrial furnaces account for a total of $6 \cdot 4 \times 10^6$ t of sulphur dioxide discharged into the air, most of it finding its way into the lungs of people living there (13). Emission of sulphur dioxide is increasing daily and will more than double in twenty years' time and, probably, treble by 2000 AD (14). The deleterious substances include sulphur dioxide, sulphur trioxide, hydrogen sulphide, carbon dioxide, carbon monoxide, nitrogen dioxide, nitrogen trioxide, ozone, ammonia, aldehydes, and acids as well as toxic metals, and carcinogenics such as 3, 4-benzopyrene (15). The range is being augmented daily.

1

Perhaps the best indication of the trend of increasing waste is given by the growth rate of domestic refuse, for which no restrictions are indicated by economic considerations. The canning of food and of a multitude of other products yields mountains of nearly indestructible refuse, such as metal tins and cans, plastic containers, also to a smaller degree, objects made of glass. Although the population, for instance in England, has grown in numbers only by a few per cent in the last thirty years, the quantity of domestic refuse has trebled in the same period. It is obvious that tipping, i.e., the storage of refuse, can no longer be indulged in, and drastic methods of refuse destruction must be invented and used. The production of undesirable smells as byproducts of this is inevitable.

The steady rise in technological output is governed by two factors essentially new in the history of mankind. The one is the increasing level of social affluence, the other, the global population explosion: daily, the birth registers of the world add 167 000 people to their lists. This increase will result in 7 000 000 000 people being alive by the turn of the millenium, i.e. twice as many as populate the earth today (16). Most of the increase will occur in this 'new world' and will involve the inhabitants of what, today, may correctly be called underdeveloped countries. This is not only so because limitation of births seems much less probable in these countries, but also because 72 % of the world population live there.

Table 1   *Factors of population increase for the year 2000*

	(1960 = 1·0)
Latin America	3·6
Africa	over 3
South Asia	over 3
East Asia	2·3
North America	about 2
USSR	about 2
Europe	1·3

Table 2 *Major regional population forecasts 1960–2000 (millions)*

	1970	1980	1990	2000
Latin America	282	374	488	624
Africa	346	449	587	768
South Asia	1090	1366	1677	2023
East Asia	910	1038	1163	1284
North America	227	262	306	354
USSR	246	278	316	353
Europe	454	479	504	527

The figures in Table 1 assume recent trends to continue to the turn of the century, whereas in Table 2 are listed median numbers of people alive in any one of the next five decades. Table 3 shows, as a comparison, world total population calculated once on the assumption that recent trends will continue, then as the medium figure which, experts in demography agree, is the more likely one to apply in fact.

Table 3 *World total population forecast (millions)*

	1970	1980	1990	2000
Recent trends continuing	3626	4487	5704	7410
Most likely actual figure	3574	4269	5068	5965

The steadily increasing rate of population growth during the last 250 years can be ascribed to the advent of mechanical industry. In addition to improved economic conditions, medical research has reduced the death rate from sickness and improved the life span of the healthy individual.

Table 4 *Historical trend of world population*

Period or Year	Duration (years)	Population (million)
7000 BC	—	10
4000	3000	25
800	3200	100
1	800	160
1200 AD	1200	350
1500	300	500
1600	100	600
1800	200	1000
1900	100	1500
1970	70	> 3000
2000	30	> 6000

That a population explosion in the next thirty years is an inevitable consequence of affluence caused by science is not accepted unanimously. DR DONALD J. BROGUE, head of the Family and Community Centre at the University of Chicago, thinks that the absolute peak population figure for the United States will be 220 000 000, and that it is most unlikely that a further increase need be anticipated, since the number of births in the USA is beginning to decline. He also notices the same tendency in other countries: in Pakistan, India, South Korea, even in China, and in some of the Latin-American countries. DR BROGUE makes the oral contraceptive, 'the Pill', responsible for this trend, and expresses the hope that a controlled birth rate will have established itself long before the year 2000. But if the present rate of increase remains steady, DESMOND MORRIS, the zoologist, calculates that by 2200 AD there will be 10 000 individuals populating every available square mile of the earth's solid surface (147).

The average expectation of life has nearly doubled in the last hundred years, and gerontologists speak of pushing the already high limit higher still. Correspondingly, production will rise with the number of people clamouring for goods. The volume of waste will probably increase out of proportion to the magnitude of production. The amount of solid waste in the USA is some 450 million kg/day or slightly more than 2 kg/person/day (17). It may have multiplied nine times by the end of the century.

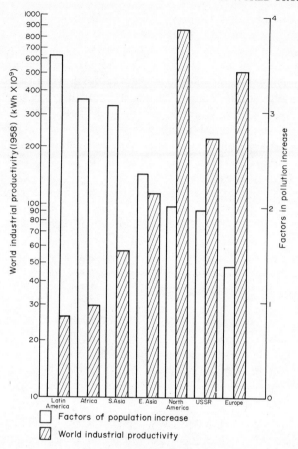

Figure 1 Relationship between rise in world population and world industrial productivity, expressed as consumption of electric power.

When the figures showing the probable rate of increase of population and of industrial activity are depicted graphically, a most interesting phenomenon emerges, Figure 1. The total annual consumption of energy, expressed in 10^9 kWh units, has been adopted as an indicator of industrial production. The two sets of data in Figure 1 show opposite tendencies. Where industrial production is lowest, human reproduction is at its peak. The converse is also true. There is one inconsistency, the USSR. Generation of energy will in the next four decades be greatly improved in Russia and when the year 2000 is approached, production of energy in the USSR will be on a par with Europe, at least.

Hand in hand with population increase will go a rise in production and it is not at all safe to predict that the goods required will be produced and supplied only by the three industrial giants: North America, Russia, and Europe. Whilst South America shows no signs of going industrial, the newly emerged African countries offer a great potential, and East Asia is, of course, a most serious competitor, with its industrial power almost completely centered on Japan. The specific consumption in that country, expressed as kWh/year per installed kW is the same as that for North America and well above the world average.

Table 5 *Specific consumption of energy*

Country	kWh/year/kW
North America	4600
USSR	4370
Europe	3900
Japan	4600
World average	4100

Japan's future as an industrial world power is supported by an economy characteristic of the Far East (Table 6). The concentration of manu-

Table 6 *Average national income per capita (18)*

North America	1750
Western Europe	940
Japan	300

facturing power and know-how in only a few centres makes for good economy, and this includes steady reduction of the proportion of waste. The 7 000 000 000 alive by 2000 AD will produce and consume goods at twice or more the rate of today.

Table 7 *Social (industrial and economic) potential*

Country	Gross social product (1967) $\times 10^6$ \$	Population (1967) $\times 10^6$	Social potential (1967) \$ per capita
United States of America	784 000	200	3920
Federal German Republic	120 000	60	2000
United Kingdom	105 000	50	2100
USSR	350 000	236	1480
Japan	112 000	100	1120
China	100 000	700	140

The last column of Table 7 suggests a top group (first three entries) at social potentials greater than \$1500 per head of population. On the first four entries, only the top three countries are likely to become exporters on a global scale of 'exported' goods and services in the widest sense. Political prestige is completely discounted here, although this may be as strong a guiding motif as are economic considerations. Even Japan is not likely to appear for some time yet as exporter in the above comprehensive sense. She will be more desirous to export her people than her goods, although recently her rate of rise in exports and production was leading the world.

While the intensity of pollution may not rise in the proportions suggested in any one place in the 'old' industrial world, twice the number of people will be there to inhale the total amount of atmospheric contaminants, which will have doubled at least. The overall picture shows that air pollution is spreading enormously in quantity and intensity, and affecting more people every day.

In underdeveloped countries future industrialization will depend on three factors: availability of capital, of raw materials, and of labour. The last item will cause a considerable upheaval in social structure similar to that experienced in Western Europe towards the close of the 18th century. The native agricultural population will flock into towns existing already or newly built, drawn by the lure of a steady guaranteed income. None of them will escape the effects of atmospheric pollution, effects which will be the more drastic and detrimental for two or three generations until the native population will become properly urbanized, i.e., more or less acclimatized to the ravages of industrially polluted air.

It would be a mistake to assume, as so often is done, that the natives of non-European countries are less sensitive to osmic stimuli on the grounds that their immediate environment, their huts or houses, often are engulfed by a cloud of smells offensive to the nose of the visitor from abroad. The only decisive factors are different levels of hygiene and sanitary conditions, and different spices and condiments used in preparing food.

Odour is a relative experience in that an osmic stimulation is evaluated against the observer's standard environment. As the native population going to work in 'new' world factories will breathe industrial odours they have not experienced before, they will be as sensitive to them as any worker in the 'old' world factories. It will be therefore a shrewd policy to consider the abatement of osmic air pollution in factories built anywhere in the world.

The international implications were brought to a painful focus by the President of the Swedish Academy of Science, PROF. S. BROHULT who, on the occasion of his traditional lecture at the annual convention of the Academy, stated that Sweden is in the immediate danger of being drowned under one million tons of sulphuric acid per year. This stupendous amount does not originate from Sweden's own chimneys, but is transported by the constant west and south-western winds from the Continent, especially from England and the great industrial centres of France, West Germany, Holland, and Belgium. PROF. BROHULT also pointed out that the mean pH value of the sea around the west coast of Sweden was as low as 3·2 during the summer. This statement shows that all attempts at getting rid of local air pollution by building higher chimneys or using other means do not get to the root of the problem, but merely shift the responsibility on to other shoulders. Whereas it is possible to deal with air pollution at its source, this is impossible when it invades another part of the globe through the free atmosphere. Global legislation is needed.

2 Some hypotheses of odour

As with all other senses, a scheme can be worked out to show how olfaction functions, Figure 2. The osmogenic cause—disregarding for the moment a more exact or detailed specification as to its nature—is inhaled through the nostrils, moves along the nasal passage ways, and stimulates the osmoceptors which react by producing specific bioelectric discharges.

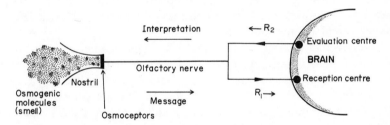

Figure 2   Block diagram representing the principle of olfaction. Osmogenic molecules (the smell) enter the nasal cavity by the nostrils and excite the osmoceptors. The specific message travels in the form of specific electric discharges along the myelinated nerve fibres to the receptor centres in the brain. After processing the message it is delivered to the evaluation centres in the brain where the olfactory information is compared with stored experience (the memory). The interpretation of the message is then relayed back along other nerve paths. There seems to be a nervous mechanism in the brain acting like a rectifier. Incoming messages can pass into the brain (to the receptor centres) only along one such nervous rectifier, R_1, and outgoing impulses from the evaluation centre along another, R_2.

These travel along the olfactory nerve and reach the lower or reception centres in the brain which transfer any reaction to the higher or evaluation centres. The result of evaluation, i.e., the effect of the primary cause,

9

seems to be projected back to the osmoceptor cells so that it seems to the observer that it is his nose (his osmoceptors) which has received and evaluated an odour. This is in agreement with the working of the other senses and with the fact that interpretation of sensory stimuli exists only in the brain and is, therefore, purely subjective.

In Figure 2, the letters R_1 and R_2 symbolize a kind of rectifier effect in that bioelectric pulses caused by external stimuli can travel towards the brain only via R_1 (afferent path), but the evaluation travels from the brain in an outward direction only via R_2 (efferent path).

Beyond this general scheme, nothing very definite can be said about the type of theory which will explain olfaction. A great many details are confirmed of how certain, if not all, elements of olfaction work individually and harmoniously together, but fundamental further questions have not been answered yet.

Today, two hypotheses hold the field, the one maintaining that chemical reactions are the cause of olfactory stimulation, the other pointing out that it is physical causes which set the olfactory nervous process into operation. In each group, several variants have been proposed and although these contrasting opinions are not really a ventilating engineer's dilemma, it is important for him to be knowledgeable about current assumptions.

The view that osmogenic stimulation is *chemical* in nature, is the older of the two. MONCRIEFF (19) has reviewed most chemical hypotheses and reveals many interesting details of this aspect of olfaction. During the last twenty years, AMOORE has proposed a new approach by assuming the olfactory stimulus to be essentially a matter of molecular shape. One of the earliest mentions of molecular morphology as the fundamental datum of odour discrimination is found in a paper by JONES and PYMAN (20). They assumed that the shape of the whole molecule rather than the structural form of its side chains was the decisive factor. EMIL FISCHER (1852–1919), a German biochemist, proposed the lock-and-key interaction of molecules, and PAUL EHRLICH (1854–1915), also a biochemist, developed, from FISCHER's principles, the side-chain theory named after him.

MONCRIEFF (*ibid.* p. 395) confirms the shape hypothesis by writing, in 1949, '. . . it seems likely that to be odorous, a substance must have molecules of prescribed shapes which will fit on certain available molecular sites in the olfactory receptors'.

AMOORE has considerably enlarged upon the basic principle and published a first account of it in 1952 (21). The current status of his steric hypothesis of odour has been discussed at length in a recent paper (22).

It is assumed that the hyperfine organization of the ultimate odour-elements in the olfactory epithelium has a surface structure which is best described as pockmarked or pitted. The shapes of these depressions are not haphazard, but show certain forms of a defined regularity. Those molecules, whose overall configurations happen to be identical with, or similar to, the shape of certain of these pits, will fit wholly, or partly, into the appropriate depression in the epithelium. This is known as 'lock-and-key' organization, and results in the molecules being so close to the osmoceptors that a sensation of odour will be produced. The character of the odour is determined—one must assume as a logical conclusion—not by the shape of the molecule, but by the shape of the minute depression. This conjecture, if found to be correct, would then account for the fact that some chemically quite unrelated substances may evoke identical, or very similar, odours; for instance Camphor $C_{10}H_{16}O$; Silicononyl alcohol $Si(C_2H_5)_3 . C_2H_4OH$ and Durene $C_6H_2(CH_3)_4$. Another camphoraceous combination is Hexachloroethane C_6Cl_6 and Trinitroacetonitrile $(NO_2)_3CCH$.

Some people may have more pits of one form than of another, or a certain shape of pit may be missing from a particular epithelium altogether. These persons will have olfactory sensations at variance with those of people having a normal distribution pattern of sensory pits. These people are called smell-blind, i.e., they are 'blind' to one or, perhaps, two definite smells.

AMOORE has allocated definite shapes to molecules of definite odours thus: Camphoraceous: *Sphere*; Musky: *Disc*; Floral: *Diamond*; Pepperminty: *Wedge*.

The *physical* hypothesis is based on the fact that the molecules of matter are in continuous motion which has different modes. They are known to the physicist as rotational oscillations and vibrational oscillations, both producing radiations the wavelength of which are characteristic of, and different for, each mode and each molecule. Values of these wavelengths have been calculated and show satisfactory coincidence with experimental results.

A mechanism is now wanted which will show how molecular radiations interact with the biological structures in the olfactory epithelium. One of the most significant discoveries in biology in recent years demonstrated odour-sensitive elements which, at one time, were thought to be the ultimate structure, yet displayed a hyperfine organization of the minutest cavities from the very heart of which a hair emerged reaching just above the surface of that cavity, Figure 3. To the physicist this structure looks like a microminiaturized magnetron (an electronic valve

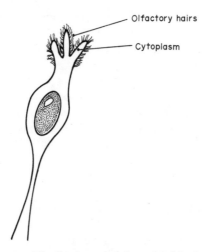

Figure 3   An osmoceptor. The fundamental shape is a bipolar structure, the 'root' joining similar processes, the 'head' forming into a number of branches each thickly covered with olfactory 'hairs'.

having a cavity for resonating operation in the microwave range). Calculations have proved that it would be resonating at about $10\mu m$ wave length. It is certainly no coincidence that this 'receiver' is tuned to the same range of wavelengths which is emitted by the molecules of a variety of substances considered to be osmogenic.

GRANT, of the University of Queensland, has first described anatomical structures in insects, called 'pit pegs', and sees in these elements a biological waveguide which, whether biological or purely physical, is essential for the propagation of long infra-red rays and microwaves. The actual biological waveguide is the sensory hair, or spine, and the pit peg is the detector, radiator (emitter), and resonator (289). The emitter best suited to infra-red wavelengths assumed to be characteristic of insects 'sending a message' is the leaky or perforated waveguide aerial in which series of slots are arranged in an axial direction along the waveguide tube. The radiation is emitted through these apertures. Exactly such structures have recently been discovered in electro-microscopic studies of insect spines and hairs, and they seem to serve the same purpose as the aerials engineered by man.

Close proximity of receiver and emitter of molecular radiation is essential in mammals, including man, because of the arrangement of the receiver, i.e., the olfactory sensor, in the depth of the nasal cavity. In animals, for instance insects, which carry their receivers on the antennae where they are freely exposed to radiation coming from all directions and at all intensities, such a proximity is inessential.

The orientation of the nostril with regard to the mouth reveals a good deal about the importance of the nose to that particular mouth—and individual. In animals which rely on olfaction as a vital sense ('macrosmatic animals'), the plane of the nostril is perpendicular to the direction of motion. There is least flow resistance and loss of acuity when sensing for prey or predators, or when digging with the snout for food, as pigs do. In the higher apes, i.e. those closest to man on the evolutionary scale, the nostrils are at an angle to the direction of forward movement, with the angle growing smaller the closer *homo sapiens* (modern man) is approached along the line of evolution. In modern man, this angle has practically been reduced to zero, so that the plane of his nostril is parallel to the direction of forward movement.

Man cannot use his nose in the same way as the dog. When the dog takes in airborne scent he holds his head in its natural position because it is then that the full cross-section of the nostril is offered to the air stream carrying the scent. If man wants to sniff the air he automatically raises his chin so that the greatest possible cross-section of his nostrils is put across the air stream. If man could bend back his head at right angles to the rest of his body, the nostrils would be able to receive the maximum volume of air, i.e., the maximum of olfactory stimulation. As things are, man's nose will only obtain maximum information from objects which he can put into his mouth. To him, the aromas from the food he eats, and from the beverages he drinks, have become more important than any other olfactory messages.

Physical hypotheses of odour have one thing in common: that they all assume radiation to be the cause of olfaction. Indeed, it would be difficult to conceive of any other physical agent to interact with living tissue, especially within the dimensions, and under the conditions of function operative in osmoception. Whether it is a range of wavelengths (290), or intramolecular processes like the oscillation of valence electrons (292), or the absorption of radiation from osmoceptors by the osmogenic molecules (293, 294), it is always radiation which is the presumed cause of olfaction. A considerable step forward in this direction has been made by WRIGHT (295, 296), and DYSON (297, 298) has attempted exact specifications of wavelength.

Molecules emit radiations at characteristic wavelengths, within a region of the medium infra-red spectrum. Depending on the structure of the molecule, the specific frequencies or wave numbers vary. The higher the grade of multiple bonds between atoms, the higher is the wave number, i.e., the shorter the wavelength. This is shown in Table 8.

Table 8 *Characteristic specific frequencies of chemical*
linkages (Adapted from GLASSTONE (299))

Bond	Specific frequency cm^{-1}	Emitted wavelength μm
C—C	800– 860	12·5 –11·6
C=C	1600–1650	6·25– 6·1
C≡C	2100–2250	4·75– 4·45
C—N	880– 930	11·4 –10·8
C=N	1650	6·1
C≡N	2150	4·66
C—O	820– 880	12·2 –11·4
C=O	1710–1750	5·8 – 5·7
C≡O	2160	4·65

If one atom in a chemical link is much heavier than the carbon atom the frequency of radiation is reduced, i.e., the wavelength increases. It is, one might imagine, more difficult for the large atom, for instance chlorine (Cl ... 35·5) to follow the vibrations of the carbon atom (C ... 12). The specific frequency of the C–Cl bond is only 650–710 cm^{-1}, and the wavelength correspondingly 15·4–14·1 μm.

Radiation of wavelength λ_i striking a molecule will be scattered and suffer a change in wavelength to λ_s. The difference between incident wavelength λ_i and scattered wavelength λ_s is called RAMAN shift (300) and is independent of the absolute values of both waves, thus

$$1/\Delta\lambda = 1/\lambda_i - 1/\lambda_s$$

and, substituting wavenumbers $\bar{v} = 1/\lambda$ the RAMAN shift can be written $\Delta\bar{v} = \bar{v}_i - \bar{v}_s$.

Osmogenic substances have characteristic RAMAN shifts which fit well into the range $1500 < \Delta\bar{v} < 3500$ so that the RAMAN spectrum extends

from about 7 μm to 3 μm, or less. Some typical RAMAN shifts are: aldehydes 1700, acetylenes 2100, sulphydryls 2500, aromatic hydrocarbons
3000.

DYSON has observed that RAMAN shifts below 1500 and beyond 3000 are
indicative of odourless substances. These findings are of importance
when considered together with anatomical data about the fine structure
of osmoceptors. These are in the form of bipolar elongated cells, the one
end being split into a number of cytoplasmic threads each of which is beset
with thousands of fine nerve fibres termed olfactory hairs, Figure 3. The
point of interest is the average length of these 'hairs'. From DYSON's
statement follows that osmogenic substances are those which have
RAMAN shifts from between 3 μm and 7 μm. The olfactory hairs,
when considered as simple antennae, can accept radiation in the RAMAN
spectrum. The shortest length of olfactory hairs was measured as 2·5 μm.
It follows that, normally, substances with RAMAN shifts outside the range
1500–3500 cannot cause an odour which is the same as saying that they
are odourless, because the olfactory hairs are either too long, or too
short to be able to tune themselves in to the incident radiation which
is outside the RAMAN spectrum. It seems that stimulation by an 'odour'
will occur on the condition that the RAMAN shift characteristic of the
molecular structure of the odour is within the defined range, and that a
certain minimum number of such molecules (a minimum radiation
energy) is present in close vicinity of the olfactory hairs to satisfy the
requirement of threshold perception.

Confirmation comes from investigations by DUANE and TYLOR (301)
who have worked with Saturnid moths. The female emits a definite pattern
of infra-red radiation between 3 μm and 11 μm corresponding to
100 THz–27·3 THz (THz = terahertz = 10^{12} Hz = 10^{12} c/s) to attract
the male Saturnid. The male insect carries an array of fine hairs on its
antenna which, on the average, are 4 μm long, or multiples thereof.
Since radiation is only emitted by sexually mature moths it is assumed
that it is caused by hormonal stimulation.

Since the infra-red wavelengths of 'odour radiation' are in the range
of from 1–15 μm it is clear that any anatomical structures acting as
physical elements in the process of transmission must be of similar, or
smaller, geometrical dimensions and are not easily discovered by
ordinary optical microscopy. GRANT (302) has discovered numerous
cavities in the body of the honeybee which, together with their associated
structures, might well be interpreted as resonant cavities of the magnetron type, especially as no other function could be found to suit these
cavities. Certain filamental elements reach into the cavity and are

interpreted as waveguides. They have an average diameter of 8 μm, and the 'micromagnetrons' could resonate at wavelengths centering at about 10 μm. This may seem a fantastic conclusion at first, but when a spectrograph of bee's honey shows a maximum emission band centering at 10 μm, it is difficult to interpret this as a chance coincidence. Experimental investigations have also shown that an electrostatic field exists between the thorax of the bee (negative) and its head with antennae (positive) the magnitude of which is in the region of 2 volt. The cavities with their filaments or 'hairs' may then correctly be interpreted as 2-V micromagnetrons.

There are many more functional details known of anatomical structures which strongly suggest the physical nature of olfaction. The most recent physical hypothesis of odour comes from Gujarat University, India, where SHAH and his co-workers (145) propose the nature of the chemical bonds in a molecule as the basic factor.

SHAH points out that in water, H—O—H, and hydrogen sulphide, H—S—H, the interatomic bonds are different. The electrons forming the bonds of the water molecule are held closer to the oxygen atom than to the two hydrogen atoms, and the bond is said to have an ionic character. In hydrogen sulphide, the electrons are more equally shared between the sulphur atom and the two hydrogens. SHAH also suggests that osmogenicity may be related to, or caused by, the presence of a bond formed by delocalized pi electrons. This bond is formed over a system of atoms when the energy relations predispose the electrons to leave their parent atoms and move into orbits that encompass the whole system of atoms. Delocalized electron bonds are found typically in aromatic molecules. Many aromatic compounds that are not odorous contain bonds with a measure of ionic character like those of water. SHAH suggests that two criteria for at least one class of osmogenic molecules are (a) possession of delocalized electrons by the molecules and (b) absence of ionic character in any of their chemical bonds.

Just as osmogenic molecules must be brought into the upper reaches of the nasal cavity in order to elicit a sensation of smell, they must be expelled from there if the observer wishes to clear his nose of one smell, and inhale another. This is difficult to achieve if, in accordance with chemical hypotheses, the molecules must go into solution, or must be adsorbed to the mucous membrane. But the physical effect of radiation can take place in an osmoceptor while the emitting molecule moves close to the nervous structure, i.e. within fractions of a micron, but without actual contact. Therefore, the strong forces of adhesion which would act in the chemical case, do not come into operation. A slight, sharp exhalation will remove the freely floating and mobile molecules just as a short sniff has brought them into the upper turbinate region.

Another physical requirement is the presence of mucus, as everybody knows who has suffered from a severe head cold. A dry membrane will not support the proper function of the nervous structures in the nose and olfaction is diminished or completely inhibited until the flow of mucus is restored.

It is not possible to verify any of the hypotheses completely, at present, and elevate them to the status of a theory. Each of the proposed hypotheses explains most of the known facts of odour sensation, but not all.

One of the more important drawbacks in present work is the lack of an exact *terminology* concerning odour. Degrees of intensity of smell, names of smells, and other classifying details vary from investigator to investigator, and a comparison of values obtained from different laboratories is often a difficult and complex matter. Sensations of sound or of sight can be exactly defined, because they can be exactly measured. But this in the study of olfaction has, at present, not succeeded. True, all chemical and physical characteristics of an odour can be measured by standard or specially developed techniques, but this does not mean that any of these measurements allow of an organoleptic interpretation. The mental background of the person experiencing an odour supplies the higher critical centres of the brain with the decisive information. A person with a pleasant memory of a particular smell, or of an occasion connected with that smell, will classify this odour as pleasing, while another person with a different background, will take a different view.

In the study of colour sensations, classification problems have been reduced to merely two different approaches: the Ostwald Theory of Colour (mostly used in Europe), and the Munsell Theory of Colour (used elsewhere). However, a blue colour remains blue under any conditions of concentration, and to any normal observer. Not so certain smells. Trimethylamine has a fishy smell at low concentration, but changes to a pungent ammoniacal smell at higher concentrations. Hydrogen sulphide loses its characteristic smell of rotten eggs at lethal concentrations, when it causes a pleasant odour sensation. Indole at low concentrations smells fragrantly of jasmin, at higher concentration of faeces, and butyl alcohol in concentration is most unpleasant and repulsive, but gives cider its attractive flavour in low concentration.

3 Thresholds of olfaction

The threshold θ is the minimum quantity of energy required to cause stimulation in a receptor system. This might be compared with the minimum sensitivity of a measuring instrument. A single-range ammeter, for instance, having a range of, say, 10 A, will not react to a current of 1 μA flowing through it although this current is constituted of $6 \cdot 242 \times 10^{12}$ electrons. Similarly, the sensory epithelium in the nose requires from several to many millions of molecules to be present, if their message shall be recognized by the brain. Measurements carried out by a number of independent investigators have shown that it is possible to arrive at certain values which may be classed as standard lower thresholds valid for the hypothetical creature 'Average Man' also known as Normal Man, or Standard Man. Individual thresholds vary greatly. Thresholds are usually measured by admitting the test odour through only one nostril. Slightly lower threshold values are obtained when both nostrils are used simultaneously (303). In order to be sensed by man, different osmogenic materials require different minimum concentrations of molecules per inhalation. Each quality of odour has its own specific lower threshold.

It is rare in practice that an offensive odour consists of only one smell. Usually, it is a mixture of two or more. At present, when neither is measured, the value relating to the dominant component will be accepted as representative for the mixture, but in the laboratory, or test room work, consideration should be given to each component, and to the type of effect the mixture has on the observer.

Several single sources
It is convenient to express the concentration, c, of the vapour in air as a fraction of the maximum acceptable concentration (MAC), M, thus c/M, which defines three cases:

$c/M < 1$ or $c < M$ mixture is acceptable
$c/M = 1$ or $c = M$ mixture is critically acceptable
$c/M > 1$ or $c > M$ mixture not acceptable

If the effects are additive the condition should obtain

$$\sum_{x=n}^{x=1} (c_x/M_x) = 1$$

Thus, for a tertiary mixture, the formula would read

$$c_1/M_1 + c_2/M_2 + c_3/M_3 = 1$$

Instance:

Polluted air contains
$c_1 = 10$ ppm of sulphur dioxide ($M_1 = 5$)
$c_2 = 6$ ppm of hydrogen sulphide ($M_2 = 10$)
$c_3 = 8$ ppm of methyl mercaptan ($M_3 = 10$)

then

$$10/5 + 6/10 + 8/10 = 34/10 > 1$$

The MAC of the mixture is greater than unity, hence inacceptable.

If the effects are not additive, each component is investigated by itself and independent of the other.

Instance:

Polluted air contains
$c_1 = 0.15$ mg/m^3 acrolein ($M_1 = 0.25$ mg/m^3)
$c_2 = 55$ mg/m^3 carbon disulphide ($M_2 = 60$ mg/m^3)

then

$$c_1/M_1 = 0.15/0.25 = 3/5 < 1$$
$$c_2/M_2 = 55/60 = 11/12 < 1$$

Both values are below unity, hence acceptable.

One mixed source
It is assumed that each component has the same v
specified temperature. If the source is a mixtu
in various concentrations c'_1 c'_2 c'_3 . . .

where

$$\sum_{x=n}^{x=1} c'_x = c'_m$$

where c'_m is the concentration of the mixture, and each substance has its own specific MAC value, $M_1 \ M_2 \ M_3 \ldots$

then

$$c'_1/M_1 + c'_2/M_2 + c'_3/M_3 \ldots = c'_m/M_m$$

and

$$c'_1 + c'_2 + c'_3 + \ldots = c'_m.$$

Taking a three-component source consisting of

$c'_1 = 1$ part diethylamine ($M_1 = 25$ ppm)
$c'_2 = 3$ parts methyl ethyl ketone (2-butanone) ($M_2 = 200$)
$c'_3 = 2$ parts ammonia ($M_3 = 50$)

then

$$c'_1 + c'_2 + c'_3 = 1 + 3 + 2 = 6 = c'_m$$

or

$$c'_1 = 1/6 \ c'_m$$
$$c'_2 = 3/6 \ c'_m$$
$$c'_3 = 2/6 \ c'_m$$

and

$$\frac{1 \ c'_m}{6 \ M_1} + \frac{3 \ c'_m}{6 \ M_2} + \frac{2 \ c'_m}{6 \ M_3} = c'_m/M_m$$

or

$$1/6M_1 + 3/6M_2 + 2/6M_3 = 1/M_m$$

ituting the values yields

$$1/6 \times 25 + 3/6 \times 200 + 2/6 \times 50 = 1/M_m$$
$$1/150 + 3/1200 + 2/300 = 1/M_m$$
$$(8 + 3 + 8)/1200 = 19/1200 = 1/M_m$$

$$0/19 = 63 \text{ approx.}$$

Because of the simplifying assumption of equal vapour pressures for all components the solution is only approximate.

The odour effects of aqueous mixtures of components have been discussed by BAKER (224), deriving mathematical expressions for odour thresholds of mixtures, and for higher concentrations. The odour threshold of a binary mixture is

$$T_x = T_A + \ln [(1-x) + 2^D x]$$

where T_x = odour threshold of the mixture, T_A = odour threshold of component A, T_B = odour threshold of component B, D = difference between A and B, i.e., $D = T_B - T_A$, x = fraction of component B in the mixture, $T_x - T_A$ = increase of the mixture olfactory threshold over that of T_A alone.

Thresholds are not expressed here as concentrations (ppm, or mg/m^3), but as having an odour intensity index (OII) which is defined in ASTM method of test D 1292 (225) as value representing the number of times the substance must be diluted by a factor of 2 to reach threshold concentration. Thus, when a substance has to be diluted to $\frac{1}{32}$nd of its concentration ($32 = 2^5$) to reach its threshold, the odour intensity index OII = 5.

For multi-component mixtures

$$T_x = \ln \left[\sum_1^N 2^{T_i} x_i \right]$$

where T_x = OII of the multi-component mixture, T_i = OII of the pure component i, x_i = fraction of component i in the mixture, N = number of components in the mixture.

For graphs and tables the original article should be consulted.

The wide range of olfactory sensitivity is demonstrated by two extreme values listed in Table 9. In the case of methanol as many as 7800 mg must be contained in 1 m^3 of air, whereas at the other end of the scale the same quantity of vanillin could be detected in 39 000 000 000 m^3 of air. No substance, natural or synthetic, is known at present which has a threshold below that of vanillin. Under threshold conditions, man's olfactory sense covers a range of $1 : 25 \times 10^{-12}$.

Whilst physics has produced instruments which will indicate a single electron, none of the senses is capable of detecting single molecules. At its most sensitive, human olfaction will react to as few as 158×10^5 molecules of vanillin per sniff, but not to fewer. This is borne out by the well-known fact that when one sniff does not produce a sensation of smell

Table 9 *Olfactory thresholds and maximum acceptable concentrations of some air pollutants (Notes, p. 24)—continued*

No.	Pollutant	Formula	Threshold ppm	Threshold mg/m³	Max. allowable concn. ppm	Max. allowable concn. mg/m³	No.
1	Acetone	$CH_3.CO.CH_3$	320	770	1000	2400	1
2	Acrolein	$CH_2.CH.CHO$	15	35**	0·1	0·25	2
3	Allyl disulphide	$CH_2.CH.CH_2.S.S.CH_2.CH.CH_2$	1×10^{-4}	6×10^{-5}			3
4	Allyl mercaptan	$CH_2.CH.CH_2.SH$	5×10^{-5}	1.5×10^{-4}			4
5	Ammonia	NH_3	3.7×10^{-2}	2.6×10^{-2}	50	35	5
6	Amyl alcohol	$CH_3.CH_2.CH_2.CH_2.CH_2.OH$	10	35	100	360	6
7	Apiole	$CH_2.CH.CH_2.CH_3O.O.CH_2.CH_3O.C_6H$	6.3×10^{-3}	5.7×10^{-2}			7
8	Benzene	C_6H_6	60	180	25	80	8
9	i-Butanol	$CH_3.CH_3.CH.CH_2.OH$	40	120	120	360	9
10	n-Butanol	$CH_3.CH_2.CH_2OH$	11	33	100	300	10
11	i-Butylacetate	$CH_3.COO.CH_3.CH_3.CH.CH_2$	4	17	200	950	11
12	n-Butylacetate	$CH_3.COO.CH_2.CH_2.CH_2.CH_3$	7	35	150	710	12
13	n-Butylformate	$H.COO.CH_3.CH_2.CH_2.CH_2$	17	70			13
14	Butyric acid	$CH_3.CH_2.CH_2.COOH$	2.8×10^{-4}	1×10^{-6}			14
15	Camphor	$C.CO.CH_2.CH.CH_2.CH_2.C.CH_3.CH_3.CH_3$	16	100**		2	15
16	Carbon disulphide	CS_2	7·7	23	20	60	16
17	Carbontetrachloride	CCl_4	200	1260**	10	65	17
18	Chlorine	Cl	1×10^{-2}	2.9×10^{-2}	1	3	18
19	Diacetyl	$CH_3.CO.CO.CH_3$	2.5×10^{-2}	8.8×10^{-2}			19
20	1,2-Dichloroethane	$Cl.CH_2.CH_2.Cl$	110	450	50	200	20
21	Diethylketone	$C_2H_5.CO.C_2H_5$	9	33			21
22	Dimethylamine	$CH_3.NH.CH_3$	6	11	10	18	22
23	Dimethyl sulphide	$CH_3.S.CH_3$	2×10^{-2}	5.1×10^{-2}			23
24	Dioxane	$CH_2.CH_2.O.O.CH_2.CH_2$	170	620**	100	360	24
25	Ethanol	C_2H_5OH	50	93	1000	1900	25
26	Ethylacetate	$CH_3.COO.C_2H_5$	50	180	400	1400	26

Table 9 Olfactory thresholds and maximum acceptable concentrations of some air pollutants (Notes, p. 24)—continued

No.	Pollutant	Formula	Threshold		Max. allowable concn.		No.
			ppm	mg/m³	ppm	mg/m³	
27	Ethyleneglycol	$CH_2.OH.CH_2.OH$	25	90	200	740	27
28	Ethyl mercaptan	$C_2H_5.SH$	1.6×10^{-5}	4×10^{-5}	10	25	28
29	Ethyl selenide	$C_2H_5.Se.C_2H_5$	6.2×10^{-5}	3.5×10^{-4}		2×10^{-1}	29
30	Ethyl selenomercaptan	$C_2H_5.SeH$	1.8×10^{-6}	8×10^{-6}		2×10^{-1}	30
31	Ethyl sulphide	$C_2H_5.S.C_2H_5$	2.5×10^{-4}	9.2×10^{-4}			31
32	Heptane	$CH_3.CH_2.CH_2.CH_2.CH_2.CH_2.CH_3$	220	930	500	2000	32
33	Hydrogen selenide	$H.SeH$	3	10**	5×10^{-2}	2×10^{-1}	33
34	Hydrogen sulphide	$H.SH$	1.1×10^{-3}	1.5×10^{-3}	10	15	34
35	Iodoform	CHI_3	3.7×10^{-4}	6.1×10^{-3}			35
36	Ionone	$CH_3.C.CH_3.CH_2.CH_2.CH.C.CH_3.$ $CH.CH.CH.CO.CH_3$	5.9×10^{-8}	4.6×10^{-7}			36
37	Methanol	CH_3OH	5900	7800**	200	260	37
38	Methylacetate	$CH_3.COO.CH_3$	200	550	200	610	38
39	Methylenechloride	CH_2Cl_2	150	550	500	1750	39
40	Methylethylketone	$CH_3.CO.C_2H_5$	25	80	200	590	40
41	Methylformate	$H.COO.CH_3$	2000	5000**	100	250	41
42	Methyleneglycol	$CH_2(OH)_2$	60	190**	25	80	42
43	Methyl-i-butylketone	$CH_3.CH_2.CH_2.CH_2.CO.CH_3$	8	32	100	410	43
44	Methyl mercaptan	$CH_3.SH$	1.1×10^{-3}	2.2×10^{-3}	10	20	44
45	Methylpropylketone	$CH_3.CH_2.CH_2.CO.CH_3$	8	27	200	700	45
46	Musk, synthetic (see notes)	$C.CH_3.CH_3.CH_3.C_6(NO_2)_3.CH_2.CH_2$	4.2×10^{-7}	5×10^{-6}			46
47	Octane	C_8H_{18}	150	710	500	2350	47
48	Ozone (see notes)	O_3	1×10^{-1}	2×10^{-1}**	1×10^{-1}	2×10^{-1}	48
49	Petrol, heavy		30	150	500	2000	49
50	Petrol, light		800	3300**	500	2000	50
51	Phenol	$C_6H_5.OH$	3	12	5	19	51

Table 9 *Olfactory thresholds and maximum acceptable concentrations of some air pollutants*

No.	Pollutant	Formula	Threshold ppm	Threshold mg/m³	Max. allowable concn. ppm	Max. allowable concn. mg/m³	No.
52	i-Propanol	$CH_3.CH.OH.CH_3$	40	90	400	980	52
53	n-Propanol	$CH_3.CH_2.CH_2.OH$	30	80	400	980	53
54	i-Propylacetate	$CH_3.COO.CH_3.CH_3.CH$	30	140	200	840	54
55	n-Propylacetate	$CH_3.COO.CH_3.CH_2.CH_2$	20	70	200	840	55
56	Propyl mercaptan	$C_3H_7.SH$	7.5×10^{-5}	2.3×10^{-4}			56
57	Pyridine	$CH.CH.CH.CH.CH.N$	1.2×10^{-2}	4×10^{-2}	5	15	57
58	Skatole	$C_6H_4.C(CH_3)CH.NH$	7.5×10^{-8}	4×10^{-7}			58
59	Sulphur dioxide	SO_2	30	79**	5	13	59
60	Tetrachloroethylene	CCl_2CCl_2	50	320	100	670	60
61	Tetrahydrofuran	$CH_2.CH_2.CH_2.CH_2.O$	30	90	200	590	61
62	Toluene	$C_6H_5.CH_3$	40	140	200	750	62
63	1,1,1-Trichloroethane	$CH_3.CCl_3$	400	2100	200	1080	63
64	Trichloroethylene	$CH.Cl.C.Cl_2$	250	1350**	100	520	64
65	Trimethylamine	$CH_3.CH_3.CH_3.N$	4	96	25	600	65
66	Valeric acid	$CH_3.CH_3.CH.CH_2.COOH$	6.2×10^{-4}	2.6×10^{-3}			66
67	Vanillin	$CH_3O.C_6H_3(OH).CHO$	3.2×10^{-8}	2×10^{-7}			67
68	Xylene	$CH_3.C_6H_4.CH_3$	20	100	100	450	68

NOTES

ppm   parts of gas or vapour per million parts of air by volume at 25°C and 760 torr.

mg/m³ approximate number of mg of matter per m of air.

** In these cases, the maximum allowable concentration has already been exceeded when the olfactory threshold is reached.

Synthetic Musk is trinitro-*tert*.-butylxylene.

Ozone: toxicity is partly due to the presence of nitrous oxide, N_2O, since the mixture of these two gases exhibits a true synergistic effect. The maximum allowable concentration of each gas in its pure state, and taken separately, is 20 ppm; but when inhaled as a mixture, 0·1 ppm is usually taken as a limiting value (306). German hygienists (307) suggest that the maximum allowable concentration of ozone should be reduced to 0·05 ppm.

REFERENCES

LITTLE, ARTHUR D., Inc.: Report C-68988, 1968.

DALLA VALLE, J. M. and H. C. DUDLEY: US Public Health Report No. 54, 1939, p. 35.

KHIKMATULAEVA, S.: *Gig. Sanit.* **32** (June, 1967) 3-6.

MAY, J.: *Staub* **26** (1966) 385.

MINISTRY OF LABOUR: *Safety, Health and Welfare.* New Series No. 8 (HMSO, London, 1966).

MONCRIEFF, R. W.: *The Chemical Senses* (Leonard Hill, 1951).

MCCORD, C. P. and W. N. WITHERIDGE: *Odours, Physiology and Control* (McGraw-Hill, New York, 1949).

VAN NIEL, C. B., A. J. KLUYVER and H. G. DERX: *Biochem. Ztg.* **210** (1929) 234.

PATTON, S.: Amer. Chem. Soc., 131st Meeting, Miami, 1957.

PATTY, F. A.: *Industrial Hygiene and Toxicology* (Interscience, New York, 1949).

TAYLOR, E. F. and F. T. BODURTHA: *Industrial Wastes*, Aug., 1960.

ZWAARDEMAKER, H.: *Die Physiologies des Geruches,* (Engelmann, Leipzig, 1895).

(because the number of molecules per sniff did not reach threshold concentration), repeated sniffing will eventually cause olfaction by accumulation of molecules on the olfactory epithelium (Figure 4).

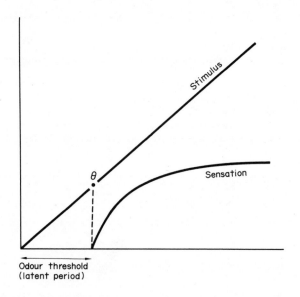

Figure 4   Graphical representation of Weber's Law, showing the odour threshold or latent period. Unless the stimulus has risen to level θ (threshold) or beyond there is no odour perception.

The values quoted in any table of osmogenic thresholds must be considered as qualitative indicators rather than as quantitative standards. When values by different research workers are studied great discrepancies may make comparisons difficult. KATZ and TALBERT (304) have pointed this out in their work giving threshold values of odours which differ from those quoted in the International Critical Tables by as much as a factor of 20. In the case of skatole, this factor is even 500. The authors explain this by the various methods used in defining the conditions of 'threshold'. Table 9 should, in fact, be restyled for engineering use by assigning relative values to the substances, for instance, trichloroethylene . . . 1, and vanillin . . . 10^9 which would mean that the human sense of smell is a thousand million times more sensitive to vanillin than to trichloroethylene; or a much less detailed schedule could be used which would still be satisfactory to the engineer.

The studies at ARTHUR D. LITTLE, Inc. (305) throw an interesting light on the effect the numerical constitution of odour panels has on results obtained. The test panel consisted of four members. Two thresholds were listed: the 50 % threshold is defined as that concentration at which 50 % of the panel members, i.e., two people, could accurately describe the type of odour. The 100 % value is the concentration at which all four members could identify the odour.

There appear to be two groups of results, the 50 % values either agreeing with the 100 % values, e.g.,

	50 %	100 %
acetaldehyde	0·21	0·21
aniline	1·00	1·00
methylene chloride	214·0	214·0

or the 50 % values being approximately one-half the 100 % threshold value, e.g.,

	50 %	100 %
acetone	46·8	100·0
carbon disulphide	0·1	0·21
toluene	2·14	4·68

Out of 56 substances (53 differing chemically from one another, and 3 being duplicated by different methods of production) about one-half (27) showed agreement between the 50 % and 100 % values, 23 being listed as having 50 % values equal to one-half the 100 % values, and 5 substances (acetic acid, benzyl chloride, dimethylformamide, ethyl acrylate, and hydrogen sulphide, prepared from Na_2S) differing in their respective value by a factor of nearly 5. Tolylene di-isocyanate has a 50 % threshold of 0·21 ppm, and a 100 % value of 2·14 ppm, i.e., ten times greater. Chromatographic purification resulted in reducing the sensitivity of one-half of the panel members:

	Odour threshold ppm	
	50 %	100 %
Tolylene di-isocyanate		
before chromatography	0·21	2·14
after chromatography	0·47	2·14

Purification removed a constituent (the 2, 6 isomer) to which the human sense of smell seems to be more sensitive than to the 2, 4 isomer which remained behind. Hence the higher threshold of the purified substance which now has values differing by a factor of 5.

The fact that there are 50% and 100% odour threshold values is a reflection on the panel, not on the substance. It indicates that large panels should always be used to arrive at a statistically significant result. In a large panel the extraordinarily gifted member, or the member who must inhale a good few sniffs to accumulate the requisite number of molecules on his olfactory epithelium before he can come to a decision about odour qualities, cannot sway the consensus of opinion of the rest of the members either way.

Concentrations in excess of an upper threshold are in some cases sensed as pain. This is so, because the trigeminal (fifth cranial) nerve has some of its endings embedded in the nasal mucosae and will react to the presence of high concentrations of odours such as ammonia, camphor, chlorine, chloroform, ether, phenol, pyridine, and others which, for this reason, are called pungent.

Table 10   *Osmogenic and nasal irritation threshold*

	Osmogenic threshold ppm	Irritation threshold ppm
Acetaldehyde	0·066	2000
Allylamine	6·2	80
Allyl mercaptan	0·000 05	150
Benzaldehyde	0·042	23
Benzyl mercaptan	0·0026	4·5
Pyridine	0·012	22 000
Thiophenol mercaptan	0·000 26	85

The ratios of osmogenic to irritation threshold of the seven substances in Table 10 cover a range from 13 to 2×10^6.

In addition to the lower threshold of osmic perception there is another characteristic value of concentration which is of importance in everyday life—the concentration of an osmogene in air at which comfort turns into discomfort.

The suggestion that lower odour thresholds be determined by measuring the olfactory limit of persons irrespective of age, sex, race, occupation, social background, and so on, should be modified when the upper or comfort threshold is discussed. This threshold is described as the eupathetic threshold, from the Greek word eupatheo which means feeling well, being comfortable.

The reason for this different approach to a seemingly identical human characteristic is in that the lower threshold is an inherent characteristic of microsmatic man, whereas the upper or eupathetic threshold is an acquired characteristic and very much dependent on upbringing, social status, past history, and nationality. There are, of course, other possibilities of grading, and a feasible classification might be according to occupation whereby the home of the person should be considered a special case of occupation.

Again, there are several ways of expressing the comfort threshold, and ppm or mg/m^3 are the most obvious choice. However, since the eupathetic threshold expresses concentrations higher than those obtaining at the osmogenic or lower threshold, the question of intensity of odour enters into it. To prevent more than one method of stipulating osmic intensity from being used, and to bring the various practices to a common denominator, it is proposed to express the eupathetic threshold as a relative figure, i.e., as a multiple of the osmic threshold. The odour intensity would then be automatically stated in the same way. The lower threshold of olfaction would be unity in each case. The eupathetic threshold of a given osmic pollutant may be one value, say 2·3, for the home, and another, say 4·6, for the place of occupation.

If there is a compound smell caused by several osmogenes, the comfort threshold for each compound should be stated since synergistic effects are not well known at present.

Other thresholds besides the osmic and eupathetic need not be considered. A limit of olfactory tolerability might come to mind, but should be discarded for the simple reason that (as noted in the preceding chapter) there are quite a number of substances to which the human sense of smell reacts differently depending on their concentration.

4 Aspects of odour

Physical aspects of odour

The sensory experience of smell is evaluated in the brain, not by the nose, and is projected back to the seat of the primary sensor (osmoceptor). This faculty is called projicience by SHERRINGTON (45). Projicience refers the sensations initiated in ourselves through the stimulation of the receptors, in this case the osmoceptors, without elaboration by any reasoned mental process, to directions and distances in the environment fairly accurately corresponding with the 'real' directions and distances of their actual sources.

For a smell to exist, the osmogenic molecules or osmogenes must enter the nasal passages in a prescribed manner, i.e., by sniffing, and interact with certain anatomical structures in the nose in order to elicit a nervous stimulation, Figure 5. Because the stimulus is registered in a special part of the brain it produces a sensation which differs from other sensory perceptions, and is characteristic of, and related to, the type of molecule interacting with the osmoceptors. A smell can therefore be described in terms of type and number of molecules entering the nose, the type relating to quality, and the number to the quantity of smell, i.e. to its intensity.

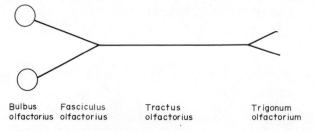

| Bulbus | Fasciculus | Tractus | Trigonum |
| olfactorius | olfactorius | olfactorius | olfactorium |

Figure 5   The basic scheme of the olfactory system. The *tractus olfactorius* (olfactory nerve) conducts the impulse to the brain.

Although it is a common expression in ventilating engineering to say that an odour is carried by the air, it is the molecules of the osmogenic matter which are carried by the air. In the absence of an observer there is no smell.

An interesting relationship between a gaseous air pollutant and the surface tension of specific media, in this case water, or water-stabilized mercury, has been discovered by EATON (33) who showed that surface tension was reduced upon introducing a pollutant. The effect was of a magnitude sufficient to be measurable only with water and water-stabilized mercury droplets. There seem to be certain analogies between the relationship: chain length of homologous alcohols and surface potential, and the relationship between the same vapours and osmogenes (34). At one time during investigations it appeared even as if this technique would lend itself to becoming a physical method of odour measurement. Investigations are also on foot to correlate physico-chemical surface interaction with olfaction. Electrical effects thus produced are expected to be proportional to physio-psychological concepts of odour, and this technique might become a first step towards a true 'electric nose'.

Although one speaks of airborne vapours as potential osmogenes, solid particulate matter in a gas phase (aerosol) or liquid particulate matter in a gaseous matrix (fog) are also classifiable as osmogenes, though of a special kind. A typical aerosol of extremely small particle size is tobacco smoke. The solid matter is not really an osmogene, but may act as a mechanical irritant when in contact with mucous membranes of the respiratory system or of the eyes. Since it is always accompanied by several tarry principles in droplet form which also exert an irritant effect, though of a chemical nature, both kinds of stimulation are usually considered together. Similarly, smoke from kilns used in the drying of fish meal, pet food, and other materials, exerts a combination of osmogenic and otherwise irritating effects on man. Solid airborne particles smaller than 1 μm are considered permanently airborne and constitute the greatest danger since they form part of man's environment. Particles larger than 1 μm have limited floating times. Their settling velocities increase with size. Liquid airborne particles are spherical when floating, but elongate when dropping, due to friction between the outermost molecules of the drop, and the air molecules.

Solid and liquid matter has a tendency to throw off its outermost molecules so that a substance is always enveloped in a cloud of molecules said to be in the gaseous state. This emission occurs at a constant rate and is a function of temperature. Volatility is a characteristic describing the number of molecules in the cloud in terms of the weight of the substance contained in 1 m³ of saturated vapour at a specified temperature. The lower the boiling point of a substance, the higher is its degree of volatility, Figure 6. The boiling point is that temperature at which the vapour pressure of the substance is in equilibrium with atmospheric pressure. If the molecules of the cloud are osmogenic, the smell of the solid substance will be perceived though, in fact, it is the molecules of the cloud (not solid, but gaseous phase) which stimulate the sense of smell.

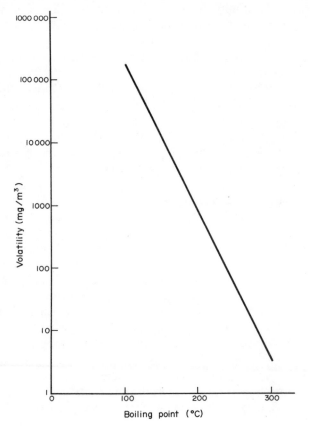

Figure 6   The effect of boiling point on volatility.

Whilst volatility may be considered a characteristic which indicates the relative ease with which a cloud of molecules forms around the mother substance (solid or liquid), vapour pressure may be thought of as indicating the energy with which the cloud tries to fly apart, i.e., the spread of the molecules, but not the actual distance.

Table 11 *Vapour pressure of osmogenes at 20°C*

Skatole	0·002 atm	Ethane	37·28 atm
Butyric acid	0·75	Acetylene	43·11
Ethylamine	1·14	Ethyl sulphide	63·82
Dimethylamine	1·66	Carbon tetrachloride	91·0
Methylamine	2·92	Acetone	184·8
Chlorine	6·57	Carbon disulphide	297·5
Ammonia	8·46	Ether	442·2
Hydrogen sulphide	17·7		

Persistence is interesting to the physicist, but harassing to the engineer. Persistence, P, is a factor expressing the time an osmogenic substance permeating free air will remain perceptible olfactorily.

$$P = (p_1/p)[\exp(1/2)(TM_1/T_1M)]$$

where p = vapour pressure, torr, M = molecular weight, T = air temperature, °K.

Indexed values refer to water as a standard

$$(p_1 = 12·7 \text{ torr}, T_1 = 288°K = 15°C, M_1 = 18, P_1 = 1).$$

Neutral values, i.e., without an index, refer to the osmogenic substance at ambient temperature t, expressed as $T°K = 273 + t°C$. The values P apply to flat (open) land. Built-up land, also single, tall objects such as houses or trees, especially poplars, will act as odour retaining surfaces. Retention is by adhesion of osmogenic molecules to the surfaces of walls, tree leaves, and other porous surfaces. Modern estate layout with wide streets and plenty of open areas between factories or houses require a doubling of the P values calculated from the above formula, but where old fashioned, narrow streets between high buildings interfere with proper ventilation of the area, and 'canyons' of brick and mortar are formed, the P values should be trebled. Similarly, multi-storey buildings cause a high P value.

The adhesion of osmogenic molecules to surfaces is the stronger, the coarser the surface. Walls of buildings, fabrics, trees, and many other surfaces forming cavities or interstices where molecules may come in extremely close contact with the substance of the surface, will hold them firmly. Whilst this retention of odours is an annoying feature in general in that it causes smell to 'hang about the place', it may, in other instances, be an intended effect directed at preventing an offensive odour from escaping from a restricted area which may be surrounded, for instance, by tall poplar trees. It is general practice with sewage works and factory estates, refineries, or single factory buildings to plant trees as a visual screen. In the case of osmogenic industries the physical adhesion of smell to the foliage is an additional advantage.

Once the molecule adheres to a surface it is very difficult to remove it again by means of ordinary cleansing methods. Washing is of little avail, and this includes rain. Large surfaces of buildings contaminated by osmogenic molecules act as secondary sources of smell when solar radiation strikes the walls and energizes some of the molecules sufficient to break away. No release of molecules takes effect after a rain or, even after washing down a building, unless with steam jets which also remove the top layer of the surface finish.

Chemical aspects

Whilst the physicist is mostly concerned with the entire molecule, the chemist will be more interested in that part of a molecule which causes its perception by the sense of smell. There are several radicals which can be detected only by olfaction, and are therefore called osmogenes, or osmophores (with reference to chromophores). Those occurring mostly in industrial smells, are

aldehydes	$-CHO$	carboxyls	$-COOH$
carbinols	$-CH_2OH$	hydroxyls	$-OH$
carbonyls	$=CO$	sulphydryls	$-SH$

Osmophores must not be thought of as units responsible for a particular smell. It depends also on the rest of the molecule, and on structural characteristics (*iso*-compounds). Different arrangements strongly affect the odour of a compound, and the same osmophore may give rise to totally different odours if in combination with other radicals. This is instanced by alcohols:

C_2H_5OH	ethyl alcohol	*sweetish odour*
C_3H_5OH	allyl alcohol	*irritating odour*
$C_9H_{19}OH$	nonyl alcohol	*offensive odour*

or by the sulphydryls:

$H.SH$	hydrogen sulphide	*rotten eggs*
$CH_3.SH$	methyl mercaptan	*'tom cat' smell*
$C_2H_5.SH$	ethyl mercaptan	*sweetish 'tom cat' smell*

Structural differences also affect odour, for instance,

| C_2H_5SCN | ethyl thiocyanate | *onions* |
| C_2H_5NCS | ethyl isothiocyanate | *mustard* |

In the sulphydryls, an alkyl radical (C_nH_{2n+1}) is substituted for the one hydrogen atom in hydrogen sulphide. Although two immediately consecutive members of the series differ by the same group of atoms $(-CH_2)$ the odours range over a wide variety, and there are many instances where, in terms of human olfaction, a relatively small change of chemical structure may swing the verdict of the observer from 'agreeable' to 'highly objectionable'. It is this individual response to changes in physical and chemical characteristics of the osmogenic molecule which introduces the many discrepancies, and even diametrically opposite views, into a discussion concerning odour. The physical and chemical aspects of smell refer to the stimulating agent, i.e. the osmogenic molecule, whereas the physiological and psychological aspects refer to the recipient of the stimulus, i.e. to the human observer. It is only by properly understanding these four fundamentally different approaches to any problem of odour, and by considering their effects together, that a useful solution can be found.

Physiological aspects

It should be clearly understood that the sense of smell is a specialized faculty of the respiratory system, and since breathing is an essential function just as is circulation, the sense of smell must be considered pre-eminent and the oldest and first of the senses. This statement is borne out by the arrangement of the cranial nerves of which there are twelve pairs in man. The sensory function of some of the nerves are given in Table 12.

Table 12   *Sensory functions of cranial nerves*

Olfaction	First cranial nerve
Vision	Second cranial nerve
Pain	Fifth cranial nerve (Trigeminus)
Hearing and equilibrium	Eighth cranial nerve
Taste	Ninth cranial nerve

Olfaction is served by the oldest, the first, pair of cranial nerves. The nasal mucosa, which is the name given to the macrostructure of the olfactory sensors, is also innervated by the fifth cranial nerve, the Trigeminus, some fibres of which are embedded in the mucosa. The upper threshold of olfaction may be identified with the pain threshold of that particular branch of the trigeminal nerve, cf. Table 10, and it is the latter which takes over when the olfactory nerve is in the refractory condition owing to over-stimulation. With many pungent odours it will be found that their pungency remains unaltered even when concentrations are very low, and the sensation of pain remains. This is so, because the trigeminal nerve is responding to stimulation, not the olfactory nerve. Since the nerve endings of both nerves are in the nasal mucosa, the sensation of both smell and pain, is referred back to the nose and hence the common experience that substances (e.g. ammonia) have an odour which causes irritation. The gap separating olfaction from nasal irritation varies greatly. In the case of allylamine, 13 times the concentration at the threshold of osmic perception will already cause a sensation of strong nasal pain. Yet, thiophenol mercaptan must be inhaled in 33 000 times the concentration of the olfactory threshold before nasal irritation will occur. There is no natural law connecting concentrations at osmic and at pain thresholds. The spread of values is random. This is so, because two different cranial nerves are concerned with these effects.

Very often, olfaction and trigeminal sensation are stimulated simultaneously so that two messages—odour and pain—reach the brain at the same time. The messages travel on a direct route to the brain since the cranial nerves are directly connected with it. Stimulation of the Trigeminus overrides any simultaneous olfactory perception, and the perception of pain prevails. This is a reflex action designed to save the respiratory system from possible damage. Trigeminal stimulation will inhibit breathing thus forcing the person to flee the dangerous atmosphere.

Location of a source of odour by a trained observer is possible within an angle of acceptance of only 7–10 deg. This is a remarkable feat considering the small distance between the nares, about 20 mm, which forms the basis on which 'olfactory triangulation' is carried out (107).

Since the olfactory epithelium, which is the seat of the osmoceptors, is situated near the root of the nose and the air passage to the epithelium is curved and long owing to the presence of three turbinate bones in the nasal cavity, air normally inhaled passes only through the lower sections of the nose and does not reach the olfactory epithelium. To get it to stream over and across the epithelium it must be given a higher speed, by sniffing it. At normal breathing, the maximum speed of the air passing through the nose approximates 3·5 m/sec (10 ft/sec) which is wind force 2 on the Beaufort scale. This is a 'slight breeze'. Taking a sharp short sniff raises the velocity to wind force 4–5, the descriptive terms being 'moderate to fresh breeze', and sneezing blows out the air through the nostrils at about 15 m/sec (50 ft/sec), i.e., wind force 7, nearly a gale. Sneezing is therefore the best method of clearing one's nostrils from any residual odour, a point to be remembered by people sitting on an odour panel.

Psychological aspects

It is only in rare cases that two people, not colour blind, will argue whether a colour should be named, for instance, blue or yellow. This is not so with odours. People argue constantly whether a smell is offensive, or strong, or irritating, or pleasant, and so on. Discrepancies arise, because olfaction is a relative experience. A judgment of odour depends not only on the permanent or fundamental make-up, such as race and sex, but also on the temporary make-up, as age, occupation, previous history of odour experiences, a cold in the head, or even on such a fleeting phenomenon as menstruation, a mental upset, and others. No wonder then, that odour panels and courts of law find it so very difficult to reach agreement between all concerned with regard to a description of the type of odour, its strength (concentration) and other technical details for which a proper definition and nomenclature have not yet been developed.

The remarkable fact that there is a group of chemical compounds which at one degree of concentration cause sensation A, at another a sensation B and, to make things worse, there are a few substances capable of producing even a third kind of osmic sensation, C, complicates things more. The two, or three, sensations are clearly different odours.

Having had an agreeable experience which, for certain reasons, has been accompanied, or caused, by a particular odour that person will always remember the odour with pleasure, even if it were mercaptan. Another person, having had a highly displeasing experience connected with roses, will scorn the smell either of that particular variety or, in extreme cases, of all roses.

Stale air, highly seasoned foods, cooking or other smells remembered faintly from childhood, may affect a person's attitude to this particular smell throughout his adult life even to the degree of making him appear peculiar to his fellows. These, and other subjective feelings towards one, or a group of odours indicate the difficulties one may have with people whose reactions to odours are important in some way and at some time. To make things worse there is no technical apparatus to measure qualitatively or quantitatively an osmic experience. Relevant data cannot be offered either for guidance or proof. On the whole, the mental approach to odours is unsatisfactory and requires much sounder foundations than can be offered at present.

5 Pathogenic effects of odour pollution

The direct effect of air pollutants on the living body need not necessarily be deleterious to health. Soot and dust, when inhaled, certainly affect the respiratory organs, but the effect could be classed as purely mechanical. Normal air contains dust, but this is quickly removed once it has reached deeper structures of the lung. The sticky mucus on the nasal mucosae holds back coarser dust particles thus preventing them from entering the remoter respiratory passages. The trachea and bronchial tubes are also coated with mucus reinforcing the cleaning effect. Mouth breathing cannot make use of the cleaning action afforded by the nasal mucosae and loses this very important line of defence. Dust collected on the mucous membranes is returned to the mouth by ciliary action and swallowed or, if returned by coughing, expectorated. Mucus transport is reversibly decreased by oxidant gases, cold, dehydration, and other traumata (96, 97). This process prevents dust from getting into, and clogging, the alveoli. The mucociliary activity of respiratory epithelia is one of the basic functions that the respiratory organism relies upon in responding to unfavourable environments, such as air pollutants, especially of the solid particulate type. However, finest dust particles will reach the deepest structures of the lung. This would, in time, become a threat to life. The very fine particles are therefore ingested by wandering cells and taken to the lymph glands which are at the root of the lungs. There, the dust is deposited (Figures 7(a) and 7(b)). The glands are always found to be black from deposited dust: significantly, more so in town dwellers than in people living in the country away from traffic and industry.

The major factor controlling the distribution of inhaled gases and particles in the alveolar air is molecular diffusion. Not more than about 15% of the inspired air is mixed mechanically during breathing. This fraction is but little affected by changes in the rate or the depth of

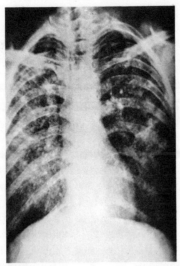

Figure 7(a)   Radiographic appearance of complicated pneumoconiosis in a 39-year old man who had worked in the graphite industry for 20 years, two of which were spent in an anthracite mill and seven in an area where products containing silica were manufactured. There are excessive changes.

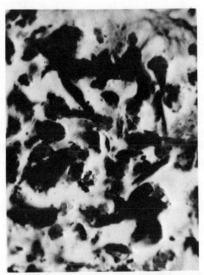

Figure 7(b)   Graphite bodies lying free in an alveolar space. The bodies consist of spicules and plates surrounded by a golden-brown substance that often forms a bulb, especially about the pointed end of each spicule. (800 ×)

breathing. From experiments it seems that the first step in pulmonary deposition of aerosols is the transfer of the particle from the tidal to the residual air. (Tidal air is the volume of air breathed in, i.e., 500 cm³ of air measured dry at 0°C, equivalent to 600 cm³ measured moist and at body temperature, 37°C. Residual air is the volume of air remaining in the lungs even after the most forcible exhalation. The lungs cannot completely be emptied of air. This volume is taken to be about 1600 cm³.) Following a single inhalation of aerosol the air expired towards the end of the cycle was free from particles (94). This means that all particles were retained in the lungs.

Pathological studies of lung tissues have revealed the effect of air pollution on the respiratory system: e.g. the amount of black dust deposited in lungs of people aged 40 and over; its weight; the percentage of silica in the lungs of persons of three age groups; and the number of dust foci of 1 mm diameter. These results establish clearly the cumulative feature of inhaling dust. These general conditions are aggravated by smoking (259, 261). The distribution of trace elements has been studied, and hypotheses have been advanced to explain it (260).

The danger from the finest particulate matter lies in its very great surface area, and that it acts as a catalyst with the molecules of airborne gaseous pollutants. A typical instance is taken from the work by FREY and CORN (99) to illustrate the point. Particles from a diesel engine are emitted in concentrations of 50 mg/m³ having a surface area of 2 m² per cubic metre of exhaust. This is about 40 m² of active surface per gram particulate matter which makes diesel exhaust a fairly powerful sorbent, at least four times as effective as various iron catalysts (60). Chemical compounds formed in the atmosphere by this effect, and solar short wave radiation, seem to be the true dangers of air pollution.

Similarly, soil will act as a catalyst and this may account for the fact that industrial fumes, when occurring in otherwise open country, seem to possess an odour of their own which differs in some ways from their characteristic smell in a fully built-up area. A first approach to this effect may have been found at the laboratories of the National Bureau of Standards where the effects of temperature and ultra-violet radiation on pyrene, $C_{16}H_{10}$, a tetracyclic hydrocarbon of coal-tar, adsorbed on garden soil, were investigated. It was discovered that these conditions produced a pyrene derivative which was not produced by pyrene under the same conditions, but adsorbed on other particulate matter (89). It is probable that other polycyclic aromatic hydrocarbons react in a similar specific way.

Liquid particulate matter (fog) behaves similar to aerosols on its way to, and through, the lungs. Investigations with radioactive iodine, ^{131}I, administered as sodium iodide of activity 100 μc in 2 ml of solution, nebulized and inhaled over a period of 15 minutes, revealed the type of distribution when detected with two balanced thallium-activated sodium iodide crystals in collimated probes (90).

The scanograms showed high concentration in the midline of the neck and thorax, principally in the mouth-pharynx space and the oeso-phagus, also in the upper abdomen, over the stomach. From this picture it becomes obvious that very little liquid particulate reached even the outer essential structures of the lungs.

This method of investigation displayed the main sites of accumulation, but made no reference to size distribution of the retained liquid particles. The scattered light particle counter solved that problem (91). Deposition in the alveoli was optimal for particles of 0·9 μm. This is confirmed by an electronmicroscope count by WATSON (95).

Table 13 relates to the case of a 75-year old man working for 18 years in a factory making carbon electrodes. In his lungs was found a solid mass 8 cm in diameter, the hollow interior (4 cm diameter) of which was filled with a black, turbid fluid.

Table 13   *Size distribution of pulmonary dust*

Size μm	Number %	Size μm	Number %
0·23	5·0	1·8	13·7
0·32	12·2	2·6	5·1
0·45	10·4	3·6	0·5
0·64	14·3	5·1	
0·90	21·2	7·2–10·2	
1·3	17·3		

Larger particles are retained preferentially in the upper respiratory tract, while submicron sized particles penetrate into the finest structures. As a general figure, retention has been measured as 30 % for particle sizes 0·2–0·4 μm, and 90 % for 3–4 μm. In heavily polluted atmospheres the high proportion of aerosols retained in the lungs not only reduces respiratory faculties, but also distends the air vesicles to such an extent as to rupture the walls separating them. This causes difficulties in expiration, not in inhalation. The disease is known as pulmonary emphysema of which, according to a recent article (92), one person in eight living in California is suffering.

In the vicinity of industrial places, there is danger of hot particles entering the lungs. Autoradiographic analysis of histologic sections revealed the presence of hot particles which show a regular inhibition of mitosis (division of active somatic cells and germ cells) in the host tissue (93). A respiratory tract dummy has been designed which serves as a model of the tract, in particular as a retention simulator. A four-stage filter simulates the human respiratory tract. The distribution of aerosols mostly found in industry was measured (102, 103).

Though air pollution is always considered an out-of-door hazard, investigations in Rotterdam have revealed the remarkable fact that indoors it sometimes exceeds that obtaining in the streets. This is a result of the backflow of chimney gases, and may constitute an important, yet little appreciated factor, in air pollution epidemiology (98).

The above data apply not only to solid particulate matter, but also to osmogenic pollutants in gaseous form. Whilst gross mechanical pollution of the air by soot, grit and smoke has been given the medical certificate of 'highly dangerous', and noise is also labelled as 'undesirable, even detrimental' to people irretrievably caught up in the bedlam of industrial production, medical opinion views air pollution by smells as nothing worse than a molestation, unless the ingredients are toxic. The law requires proof that a nuisance has been committed before action can be taken. People tend unduly to minimize the effects of malodours, and say, 'but, it's only a smell'.

Medical aspects of air pollution cannot be discussed without considering such things as local climate, general climate, and weather conditions, which affect the dispersion of fumes (temperature inversion, wind), or cause the catalytic conversion (sunshine) to disagreeable or noxious compounds.

Whilst most osmogenic fumes are non-toxic, the popular fallacy is sometimes maintained that a person's health has been ruined by 'that smell'. In lawsuits concerning osmogenic air pollution the minds of all involved are difficult to screen from preconceived ideas. Emotional aspects mingle with indeterminacies and inaccuracies of witnesses' testimonies to the despair of judge and experts, and when all other pleas have failed to impress the court it is the 'I felt thoroughly sick' or, best of all, 'baby was sick' which will have greatest impact.

A few gaseous compounds exist which may present a real hazard to man.

The danger lies not in their odour or in their toxicity alone, but in the aggravation of circumstance stemming from the fact that the maximum admissible concentration is lower than the odour threshold so that not even the nose, this most sensitive detector of smell, can perceive the danger in time.

Table 14 *Some vapours whose maximum acceptable concentrations (MAC) are lower than their olfactory threshold*

Vapour	Olfactory threshold	MAC	Ratio
Acrolein	35	0·25	1400
Camphor	100	2	50
Dioxane	620	360	1·7
Hydrogen selenide	10	0·2	50
Methanol	7800	260	30
Methylformate	5000	250	20
Methylene glycol	190	80	2·4
Ozone	0·2	0·2	1
Ozone (latest proposal)	0·2	0·05	4
Petrol, light	3300	200	16·5
Sulphur dioxide	79	13	6·1
Trichlorethylene	1350	520	2·6

(Limits in mg/m^3; ratio is threshold/MAC.)

In these, and similar, instances an instrumental detection and warning system is of paramount importance. Organoleptic assessment is hopelessly inadequate in most of the cases. If the ratio olfactory threshold/MAC is less than unity, the smell will be perceived by the nose before the pathogenic concentration is reached. For unity ratio the slightest impairment in olfactory function will raise the threshold value above MAC level, and ratios greater than unity indicate danger when relying on the nose instead of using an instrumental detector. The cases threshold/MAC < 1 and threshold/MAC > 1 are clear cut. No danger in the first, immediate danger in the latter instance. Only when the ratio equals, or approaches, unity is there hidden danger. Ozone is an instance. Several years ago, its maximum acceptable level was put at 1 ppm, by some at 0·8 ppm. Russian publications strongly suggested a reduction to 0·5 ppm and, finally, the American Conference of Governmental Industrial Hygienists standardized it at 0·1 ppm. German hygienists now accept only one half that value, 0·05 ppm. NASR has reviewed the biochemical aspects of ozone intoxication (247).

Other peculiarities of olfaction are the change-about turns in the characteristics of various osmogens. The most important from an industrial point of view is that of hydrogen sulphide, often described as sewer gas. In low concentrations it has a smell of rotten eggs. In concentrations of near lethal level this smell gives way to a quite pleasant odour (100). The absence of a warning smell has caused many a death of workmen entering a sewer or tank. Whilst it is usual to say that the osmogen changes its odour, it is certainly more correct to say that the brain reacts differently to low and high stimuli caused by the same chemical substance, here hydrogen sulphide.

The message from the stimulated receptor to the brain takes the form of electrical pulses travelling at speed along the nerve connecting the two elements. In warm-blooded animals including man this velocity is approximately 30 m/sec for medullated sensory fibres (101). Each pulse has its own characteristic frequency and an increase in stimulus is expressed by an increase in frequency, not in amplitude. A weaker stimulus will start off pulses along fewer fibres leading off the receptor. It can be argued that a low concentration of hydrogen sulphide will send a small number of pulses along the olfactory nerve to the brain where they will cause a certain characteristic reaction (sensation). At higher concentration the maximum number of pulses will be conducted along all medullated nerve fibres in the olfactory nerve at frequencies different from those in the first mentioned case. This will produce a different reaction altogether resulting in a different sensation of hydrogen sulphide. Since the pleasant smell is perceptible only at higher concentrations it is justified to assume that two types of osmoceptors are engaged in sensing hydrogen sulphide, the one group having a low odour threshold the sensation of which is interpreted as 'offensive', the other group having a higher odour threshold (thus remaining ineffective at 'offensive', i.e., lower concentrations) and reacting only at higher concentrations of hydrogen sulphide which sensation is interpreted as 'pleasant'.

The type of change, i.e., from offensive to pleasant, is purely coincidental and a subjective interpretation. The physiological function of any such odour change must rest, ultimately, with different groups of osmoceptors, having different threshold levels. This is nothing unusual. A duality of sensors and perception levels provided for the same stimulant is well known in the sense of vision. There are two types of sensory elements: the rods, reacting at very lowest levels of illumination, with the brain producing a black-and-white image of the object; and the cones, reacting to the same type of light, but at substantially higher levels of illumination, with the brain producing an image in colour. When the cones are stimulated the rods remain inoperative, and vice-versa.

Whilst it is usually agreed that smells have no deleterious effect on health unless the airborne fumes are not only osmogenic, but also toxic, it should be remembered that there are cases where the material producing offensive smells also harbours bacteria, many of them harmful, as in sewage works. Many of these micro-organisms persist for considerable lengths of time and distance from the works. Mostly, they are bacteria of proved pathogenicity in the respiratory tract and far more dangerous than the enteric pathogens also carried in the air stream. The most frequently isolated bacterium was *Klebsiella pneumoniae* and its distribution and life were of considerable magnitude. The inherent danger moment was greatly increased by the fact that viable *K. pneumoniae* travelled on aerosol particles of a size which easily penetrates to the finest lung structures carrying the micro-organism into the alveoli (126).

There is also a small group of compounds the presence of which in environmental air will increase man's tolerance to specified other compounds. This may constitute a series of hazards if the second compound is deleterious to health. An instance is ketene, $CH_2:CO$, the simplest possible ketone. CARROLL suggests (230) that it occurs in tobacco smoke and may be carcinogenic. It is a colourless gas of a very strong, sharp, acrid odour and might, perhaps, be classed as an irritant rather than an osmogen. It is a product in the pyrolysis of oxygen-containing compounds and produced in quantity in smouldering combustion together with acrolein. It can be smelt when chromatographic techniques will detect it only in concentrations many times the organoleptic threshold. Ketene increases man's tolerance to ozone. This can be dangerous to a cigarette smoker in an ozonized atmosphere. Pipe smokers are not so affected since ketene breaks down in the moisture of the pipe stem. The maximum acceptable concentration of ketene is 0.5 ppm $= 0.9$ mg/m^3.

Carbon disulphide is not only the cause of an unpleasant odour, it is also toxic to man. A dangerous exposure causes widespread and severe damage to many parts of the body. There is loss in body weight, hypertrophy of the adrenals, pathologic changes can occur in the brain by lymphocytic infiltration, and spinal cord lesions may develop (248). Affected workers in viscose staple fibre factories have recovered on withdrawal from the CS_2 area. However, chronic renal dysfunction may be a potential long-term effect (249). Where CS_2 concentration is high as it often is in the production of viscose staple fibre (150–350 mg/m^3), the excretion of zinc in urine, otherwise a fairly stable function, increases (250). Data on retention, biotransformation, and pathogenic effects of carbon disulphide are given by BRIEGER (279).

Polytetrafluoroethylene, PTFE, is a plastic which is inert to most chemicals and resists high temperatures up to, and in certain cases, above 250°C. The cold plastic has no smell at all. With rising temperature, a slight smell develops which becomes acrid at pyrolysis (at about 550°C). It then produces toxic gases, the principal toxic component being carbonylfluoride, COF_2. In a worker inhaling these fumes, changes may develop in the lungs and liver after prolonged exposure, but such cases are rare in practice unless the workers are confined to the room where pyrolysis occurs. After withdrawal in good time, pulmonary irritation may persist for a few days, but will recede (251).

6 Methods of detection and measurement

Standard methods for the detection and measurement of gaseous compounds in air are listed in every textbook on the subject. Chemical methods rank first, chronologically, but the number of elegant physical methods available for detecting the most minute quantities of a substance in air is increasing fast. Although any standard method of detecting and measuring a gas or vapour is applicable to the problems of air pollution, it is mostly the techniques which differentiate between methods suitable for a laboratory on the ground and, for instance, a laboratory on wheels or, even, on wings.

Methods peculiar to an air pollution problem include the sampling of polluted air which can be done by captive balloons stationed at different points in the area and rising to different heights; or the balloon may be mounted with its winch on a suitable vehicle and travel to the various sampling stations. If large areas are concerned, helicopters or small aircraft will be carrying highly specialized laboratories to the appropriate levels in the atmosphere, and bring back samples of air and a great deal of other information relevant to air pollution. Such an experiment has been forecast by the State of California (52).

The UK Ministry of Technology has ordered a close collaboration between the Atomic Energy Research Establishment at Harwell and the Ministry's Warren Spring Laboratory at Stevenage with regard to investigations into air pollution. The vast and highly specialized instrumentation of AERE will involve the use of chromatography, electronmicroscopy, radio-activation analysis, radio-tracers, and other techniques, in air pollution research. Also to be studied is the absorption of sulphur dioxide and other gaseous pollutants by solid or liquid particulate matter; the formation of sulphuric acid in air; lifetimes of pollutants; hazards from the inhalation of polluted air; the effect on plant life and the corrosion of building materials, and the factors controlling the wash-out of pollutants from the air by rain.

47

From time to time, new pollutants are discovered in the air, requiring exact definition and quantitative description. DUCKWORTH and KUPCHANKO (53) have devised the dosage-area coefficient which is the product of pollutant concentration in ppm, time in hours, and area in square kilometres or square miles. Although the concept of dosage-area has been developed as a smog index, there is no reason why it should not be adopted to specify an area-wide odour index, or odour intensity index.

There are definite steps required before an index could be established. First, the daily concentrations of all pollutants under investigation must be established; most efficiently, by an aircraft flying at a specified altitude taking air samples. If the area is very large, an air monitoring network should be established so that the relationship between pollutant concentrations and time of day can be monitored. Meteorological data, such as wind velocities and directions, will also be recorded.

Next, the dosage pattern is found from the graphical representation of the above findings. The graphs are called by the authors isopleths (Gk: isos = equal; pletho = to be full) as they connect points, i.e., locations, having the same concentration of pollutants.

The third step is to measure the area between isopleths either by co-ordinates or with the aid of a planimeter. The individual areas are then multiplied by their relevant dose levels.

Finally, the sum total is taken of all partial products (area by concentration). The geographic configuration is accounted for by a coefficient called the type designator.

Whilst the DUCKWORTH–KUPCHANKO method yields purely quantitative results, a method using infra-red rays is available for the very fast identification of air pollutants (54). The polluted air is deposited in pulses on a deep-cooled alkali halide substrate and is then subjected to infra-red analysis. The sample preparation and deposition requires only seconds. The infra-red spectra are well suited for digital analysis. Spectral data can be stored within a central computer which is linked via telephone lines to any number of investigators or experimental stations. These transmit spectral data to the central computer and an analysis is communicated to the local station within a few minutes.

Tracer, or radio-activation techniques have proved valuable. A tracer system has been developed by the United States Centre for Air Pollution (55) which, using sulphur hexafluoride, SF_6, permits of detecting one part substance in a hundred million parts of air (10 ppb), although this tracer is not radioactive. Its value lies in the fact that sulphur hexafluoride is normally absent from the air. The gas is released together with the pollutant and the air is then sampled in a downwind direction, and analysed for SF_6, hence revealing also the path of the pollutant.

Neutron activation of airborne particulate matter has been carried out at Harwell on trace elements (56), but is adaptable to investigations into air pollution. For the radiochemical determination of mercury vapour in air the 203-isotope of mercury is used (59). The air suspected of containing mercury vapour is passed through a solution of ^{203}Hg-labelled mercuric acetate and potassium chloride, when isotope exchange takes place. The air will now contain the same amount of mercury atoms, but each labelled and of the same specific activity as the reagent solution. The ^{203}Hg is absorbed on hopcalite (a mixture of suitable metal oxides) and the radio isotope is estimated by γ-scintillation counting. This is a simple method which can be used for long run sampling, continuous recording tests, and for one-off determinations. The standard deviation of this method is

0·004 μg of Hg per litre of air up to concentrations of 0·2 μg Hg/1
0·075 μg of Hg per litre of air for concentrations of 0·2 − 1·2 μg Hg/1

This method can be suitably varied to measure any other pollutants whether osmogenic or not.

Particulate osmogenic pollution is quantitatively determined by the simple physical methods of evaluating β-attenuation and expressing it in terms of total or specific weight of particulate matter. Sources of β-radiation are Strontium-90 or Yttrium-90. The accuracy of measurement seems to depend on the layer of dust, i.e., on the weight of dust, retained on the filter (57).

A small and easy to handle apparatus built at the UK Government Chemist's Laboratory permits of determining gas concentrations down to 1 part per hundred million parts of air, sufficient for all purposes including the detection of toxic gases (58).

The high rate of air sampling, up to 100 1/min, has been achieved by replacing the solid or liquid filters, or impregnated papers, all of which offer a high flow resistance, by an indicating powder. This may be a reactor-impregnated silica gel of 50–100 mesh (British Patent 1017545). The effect of the instrument is based on colour changes of the indicator after reacting with the pollutant. It is therefore necessary to prepare for each reagent a series of air samples of known concentration, and keep a reference library of colour scales for different pollutants and different reagents. Air samples taken in the field can be evaluated either by direct visual inspection or by any suitable type of colour comparator.

There is also a time factor involved due to accumulation of the pollutant if it is diluted by air to a high degree. The authors give several instances:

arsine in air at 0·05 ppm takes	10 min
sulphur dioxide at 0·25 ppm takes	2 min
mercury vapour at 50 $\mu g/m^3$ takes	10 min

Chemical detection tests for ozone, hydrogen sulphide, hydrogen peroxide, carbon monoxide, aldehydes, and other compounds are not usually expected of the engineer. Some relevant tests are listed in the literature (60). The General Electric Co. has produced a condensation nuclei counter (115) which is capable of counting airborne pollutants down to 0·005 μm size.

Ozone

Qualitative tests for ozone are carried out by means of test papers. Tetrabase paper is made by soaking filter paper in an alcoholic solution of tetramethyl-di-p-diamino-diphenyl methane (tetramethyl base) $(CH_3)_2N.C_6H_4.CH_2.C_6H_4.N(CH_3)_2$. It changes colour in the presence of ozone to pale violet. The halogens turn it blue. Hydrogen peroxide is not indicated. Sensitivity can be enhanced by adding a few drops of potassium acetate and dilute acetic acid. Too much of the latter will produce a non-specific change to blue in the presence of either ozone or the halogens (61).

Benzidine test paper is produced by moistening filter paper with an alcoholic saturated solution of di-p-diamino-diphenyl, known as bi-p-bianiline $(NH_2.C_6H_4.C_6H_4.NH_2)_2$. It will change colour to

brown   in the presence of ozone
blue, then brown   in the presence of chlorine
blue   in the presence of bromine, nitrogen dioxide

It remains unaffected by hydrogen peroxide, hydrogen sulphide, ammonia, ammonium sulphide, and hydrogen cyanide (62).

Quantitative methods measuring ozone use potassium iodide. WADELIN (63) liberates first iodine

$$O_3 + 2KI + H_2O \rightarrow O_2 + 2KOH + I_2$$

and then adds sodium thiosulphate which reacts with the iodine

$$2\,Na_2S_2O_3 + I_2 \rightarrow 2\,NaI + Na_2S_4O_6$$

sodium thiosulphate   sodium tetrathionate

producing sodium tetrathionate which is then titrated.

KAESS (64) has introduced a variant of WADELIN's method achieving a sensitivity of 0·05 μg ozone per cm^3 potassium iodide.

Colorimetric quantitative measurements are used, for instance, the colour change in dilute N-phenyl-2-naphthylamine dissolved in o-dichlorobenzene (65), or in sodium diphenylamine sulphonate (66). The phenylnaphthylamine method developed mainly by DELMAN (65) and patented in the United States, US Pat. 2 849 291, has been improved by GERMAN et al. (73), so that the required volume of reagent is now only 15 ml. Absorbance is measured at 4350 Å. The instrument is precisely calibrated by an iodometric determination based on WADELIN's reaction $O_3 + 2KI + H_2O \rightarrow O_2 + 2KOH + I_2$. A concentration of 0.57 μg of ozone per litre air is the lower limit of detection.

Further methods are given by BOWEN (67) and other investigators (68, 69, 70). The acid potassium iodide method (75) is unaffected by air, oxygen, temperature variations, and reducing gases in the air stream, such as hydrogen sulphide and sulphur dioxide. A chromium oxide, CrO_3, scrubber removes the reducing gases. This method is reliable and effective from 1–70 ppm of ozone. A similar technique is reported from Rumania (88) and seems to be valuable in the industrial field where oxidants and reductants occur simultaneously in the air. Nitrogen oxides, sulphur dioxide, and hydrogen sulphide are oxidized and retained on silica gel which is saturated with 0.01 M $K_2Cr_2O_7$ in H_2SO_4 of density 1.84. The ozone then reacts with $KCdI_3$ or KI and can be detected photometrically or by titration. Accuracy is 0.1 μg per 2 ml, and the precision of detection is 0.1 μg. Another method makes use of the reaction between ozone and 4, 4'-di-methoxystilbene. The resulting anisaldehyde is then detected by SAWICKI's method with a 5% solution of fluoranthene. This acts as the coupling reagent. A blue colour results and is measured by spectrophotometry. There is no interference from peroxyacetic acid, peroxyacetyl nitrate, methyl hydroperoxide, sulphur dioxide, or nitrogen dioxide (280).

KOBAYASHI and coworkers (71, 72) have developed methods of measuring the distribution of ozone in the atmosphere. Parts I and III of their paper discuss investigations into the vertical distribution of atmospheric ozone, while the middle section deals with the application of their methods to ozone at ground level. The latter is a variant on WADELIN's technique (63). The operational principle employs the reaction of ozone with a neutrally buffered potassium iodide solution which contains a known amount of sodium thiosulphate. The iodine in the solution is liberated by electrolysis. Sensitivity is about 2 μg/m^3/sec with a short response time of 30 sec. The Japanese researchers have produced instruments based on the methods outlined in their paper.

A physical method of measuring the vertical distribution of ozone uses two beams of ultra-violet radiation passing through collimators, optical solar filters, and two photocells with a-c amplifiers and integral chopping circuit. A telemetering system relays the data to stations on the ground. The dominant wavelengths and half-widths of the two beams are 3050 Å (270 Å) and 3400 Å (120 Å). The other method is chemical using a neutral buffered solution of halides with which ozone will react, liberating the halogen. This is then detected coulometrically. The instrument consists of a platinum gauze cathode and an active carbon anode. The electrodes are immersed in the solution. The electric current thus produced depends on the amount of ozone aspirated to flow between the two electrodes. The device is reported to be reliable, easy to handle, and stable in operation. The work of KOBAYASHI (72) on the vertical distribution of ozone up to 3 km altitude has been extended for satellite observations by DAVE and MATEER (87). Their preliminary results indicate that the wavelength region in which measurements should be made, lies between about 3125 Å and 3175 Å if the observations are restricted to the nadir direction. They also find that wavelengths in excess of 3175 Å offer poor sensitivity to total ozone. Accuracy of measurement is only 10%. The effects of reflectivity are reduced if the ranges 3125–3175 Å and 3300 Å are used. Accuracy rises to 5%.

Methods of chemiluminescence suffer from instability by direct oxidation. Rhodamine B has been used as the luminescent light emitter to which gallic acid in ethanol has been added as an ozone acceptor. Observations have been made with an oscillographic record of the light emitted by single bubbles of ozonized air. Another fluorophotometric method has been devised by WATANABE and NAKADAI of the Osaka Institute of Hygiene (212). A solution of 9-10-dihydroacridine in ethanol is acidified with $6N$ acetic acid and will fluoresce if oxidized by ozone. Maximum intensity of fluorescence is at 4820 Å. Between 0·1 and 0·35 μg acridine per ml, the intensity of fluorescence is directly and linearly proportional to the concentration of acridine. The method is specific with one exception: nitrogen peroxide, N_2O_4, interferes with the measurement. Since ozone produced by ultra-violet discharge tubes produces no nitrous oxides, WATANABE's method gives specific indications of ozone under these conditions of O_3 generation.

A recent and entirely physical method uses the optical analogue of radar, i.e., lidar (light detection and ranging). Pulses of laser light are used for mapping the distribution of particulate matter in the atmosphere. Lidar, because of its small wavelengths, can 'see' subvisible particles.

A new chemical method of quantitative ozone determination has been devised by NASH (74). He uses diacetyl-dihydro-lutidine as a reagent. (Dihydro-lutidine, $C_7H_{11}N$, is a ptomaine from cod-liver oil. Ptomaines are animal alkaloids from the decomposition of dead animal matter.) He reports that its sensitivity for ozone is 500 times that of the standard potassium iodide method, and at least 10 times that of determinations using thiosulphate. The reagent is prepared quite easily and the air bubbled through an aqueous solution of it. The amount of ozone in air is measured as the loss in optical density at 4120 Å. Ozone causes a shift to 3070 Å, a new strong spectral band. The compound responsible for the shift has not yet been identified, but it is certain that the shift is not caused by the oxidation product of the reagent. The reaction is specific for ozone. Sulphur dioxide, nitrogen dioxide, and peroxides do not interfere.

The ferrous ammonium thiocyanate reagent for evaluation of ozone concentrations has a very high absorptivity, about 30 000, at ozone concentrations below 100 pphm. Collection rate, and the bubbler frit size, are of importance (76).

Hydrogen peroxide
Hydrogen peroxide is detected by a physical direct, or a chemical indirect, method. The first collects a sample of air $+ H_2O_2$ by aspiration in water. Add 3 ml of 0·5 % vanadic acid to 5 ml of the solution and measure the absorption of monokymatic (filtered) light of 4400 Å wavelength 15 min after having prepared the test solution. The absorption of the pencil of light depends linearly on the concentration over the range 3–200 γ.

The chemical method requires collecting the air sample in 10 ml water, 10 ml 0·01 N ammonium iron sulphate, $(NH_4)_2Fe(SO_4)_2$. 6 H_2O and 0·025 ml 10 M potassium thiocyanate, KSCN. This reaction is non-specific for hydrogen peroxide (77).

Carbon monoxide
Carbon monoxide in air is detected and measured by the reaction of CO with hot mercuric oxide, HgO, followed by the photometric determination of the amount of mercury vapour liberated. The equipment utilizing this reaction is not affected by hydrogen or methane in air (78).

Hydrogen sulphide
Hydrogen sulphide in sewage is determined by methylene blue which becomes discoloured in the presence of H_2S or sulphur reducing bacteria. To 0·5 cm³ of a 0·10 % aqueous solution of methylene blue are added 250 cm³ sewage in a glass-stoppered bottle. No air should be left in the bottle which is stored in the dark at 20°C. Colour is checked every hour

(79). Trace quantities of hydrogen sulphide are detected by Smith's method (81). This is a modification of standard methods taking into account the simultaneous presence of hydrogen sulphide and sulphur dioxide in industrial atmospheres. A 2-hour sample of air will allow to measure H_2S concentration to 10^{-4} ppm, which is one-tenth of the olfactory threshold value.

Sulphur dioxide

A most accurate determination of sulphur dioxide, hydrogen sulphide, and methyl mercaptan in stack gases can be carried out by chromatography (80) which is the method of choice for the detection, identification, and separation of gases contained in a single small air sample.

A Mine Safety Appliance Company has produced a sensitive detector of sulphur dioxide and hydrogen sulphide mixtures in air. A known volume of gas is passed through the indicator which consists of solid matter in a transparent glass tube. The indicator for sulphur dioxide is phenol red or bromothymol blue, and silver cyanide for hydrogen sulphide. The indicators change colour when the respective gas passes through the tube, the length of colour change being proportional to the amount of SO_2 or H_2S in air. For simultaneous detection and measurement of both gases, the air is passed through the tube. The inorganic solid is, for instance, silica gel, and a layer of activated alumina, Al_2O_3, is used as the carrier of the silver cyanide with which it is impregnated. Since both indicators are specific, sulphur dioxide will cause a colour change in the dye, and hydrogen sulphide in the cyanide. Process and equipment are patented (British Patent 1 084 469, 1967).

An automatic signalling instrument for the detection and measurement of gases and vapours in air makes use of continuous colorimetry. The measuring cell is automatically filled and emptied at predetermined periods of time and optical as well as acoustical signals are given when maximum acceptable concentration is reached by a particular gas. The apparatus is used mainly for the determination of hydrogen sulphide, nitrogen peroxide, and chlorine in air.

Nitrogen oxides

The air containing nitrous oxides passes together with the reagent through a beaded column where diazotation occurs. The reagent is N-1-naphthylethylenediamine dihydrochloride. Discoloration occurs and is compared with a scale of standards either visually or photoelectrically (82).

Aldehydes

These are detected by absorbing them on silica gel impregnated with *p*-diphenylene diamine and hydrogen peroxide. Low concentrations of aldehydes can be detected and estimated. There is a specific colour change, and the intensity of colour is directly proportional to the concentration of the aldehydes. A minimum of 10^{-8} mole % can be detected (83).

Phenols

Airborne phenols in small quantities, down to 1×10^{-5} mg/l, are detected by photosensitors using *p*-nitraniline as a reagent (143). The odour threshold of phenol is 1.2×10^{-3} mg/l and the maximally acceptable concentration is 1.9×10^{-2} mg/l.

Air standards: the 'electric nose'

An attempt to produce primary air standards for exact measurements has been made by the PolyScience Corporation (262). They have produced ten liquid preparations of benzene, butane, butene-1, cis-butene-2, trans-butene-2, hydrogen peroxide, isobutylene, nitrogen dioxide, propane, sulphur dioxide. The standards are sealed in PTFE (Teflon) tubes and allow quantitative determinations in an unknown sample to the ppm level.

Several apparatuses have been constructed which, it was hoped by their designers, would fill the need for an impersonal, objective and reliable odour measuring equipment. Not many have realized the abysmal gap separating all those devices which measure chemical concentration from the one and only device in existence which reacts to—one is tempted to use an unscientific expression for lack of a better description—all those intangible stimulants which cause either aesthetic pleasure, or vulgar offence—the nose.

PROFESSOR EATON of Purdue University says of his discovery that the pendant drop technique for the measurement of surface tension may point a way to the measurement of odour (not chemical concentration): 'The results of this investigation indicate certain analogies between the changes in surface potential and the olfactory stimulation produced by certain vapours, but should not be interpreted to imply that odour has been measured by electrical-mechanical means' (213, 214).

Another method of measurement was described by SEIYAMA and KAGAWA of Kyushu University, Japan (215). They use a semiconductor thin-film adsorber to detect various organic substances in air. The detector is a thin film of zinc oxide on a silicate plate 5×30 mm, oxidized to a semiconductor condition, and about 0.5 μm thick. Optimal operating temperature is 500°C. The response of the zinc oxide film to the concentration of the samples in air is logarithmic. This is a response similar to that of the sense of smell (Weber-Fechner law). The sensitivity of the detector varies with the nature of the chemical substance offered to it for testing, and has a lower limit (threshold) of sensitivity at about 10–100 ppb. This caused reporters to write in dailies and other journals and magazines that an 'electronic nose' was just about to be discovered, and in support of this hopeful suggestion it is mentioned that the detector is most sensitive to molecules with the highest electron-donating properties. The thin-film detector's sensitivity to the following osmogens decreases from amines to cyanides:

amines > ethers > mercaptans > alcohols > ketones > aldehydes > carboxylic acids > cyanides

Mercaptans range third place, behind amines—such is the characteristic of the zinc oxide film. The human sense of smell knows a different order, placing mercaptans first, and amines very far behind. Mercaptan thresholds are of the order of 10^{-6}–10^{-5} ppm, and that of amines 10^{-1}, or higher.

The sequence of effectiveness of detection, the sensitivity, for the zinc oxide film is the correct gradation with respect to the electron donating potential of the substances quoted in the series which is a maximum for amines, and a minimum for cyanides. This is consonant with the characteristics of the zinc oxide layer which is an n-type semiconductor. Its conduction bands are capable of accepting almost any number of electrons. Molecules of longer chain length or higher dipole moment are preferentially adsorbed on the surface of the semiconductor detector which is interpreted as a higher sensitivity of the detector for substances consisting of that type of molecule.

Summarizing the tantalizing subject of electronic noses, it must be pointed out most definitely that no such instrument has yet been developed; that there is a great need for one; and that there is little hope, at present, that such an instrument will be designed in the near future. Meanwhile one must accept chemical concentration for odour interpretation or sensation, bearing in mind that there are some correlations between the two quantities, but no identity.

7 Social aspects of odour

The earliest records of human civilization suggest that a standard of cleanliness was achieved, even if early rules of hygiene were wrapped up in ritual ordinances and prescribed periodical ablutions. It would therefore seem wrong to assume that scent was originally intended, and used, for the masking of unpleasant body odours. During the Middle Ages in Europe, however, and right into the 17th and 18th centuries, perfumes were certainly employed with this object.

Today, social life includes all amenities of hygiene and sanitation, such as healthy food, reasonable clothing, bathrooms and drainage. On the other hand, the same social life congregates hundreds, even thousands of people in one space either for work (offices, factories), or for pleasure (cinemas, theatres, dance halls, indoor stadia). Inevitably, body odour becomes apparent, but even more so the odour from personal scent used by women—not only behind the ears, and by men—not only in hair creams. Even the most ingeniously designed ventilating system must sometimes capitulate before the mixture of perfume and sweat and odours of metabolic origin.

Odour has been made a great selling agent, a commodity without which modern life would be unthinkable. Not only are housewives told that their bathroom must smell of lilies of the valley, they are also told—probably on the same page of the advertisement, that they should destroy all smells in the bathroom; that they should use a deodorant against body odour, a deodorizing lozenge after eating, after drinking, after smoking, after . . . ; that they can now buy permanently odorized wallpaper, perfumed toilet paper, permanently odorized lingerie, and permanently deodorized whale meat.

In fact, the odourmania has so spread—escalated may here be the correct term—that people tend to live in environments of 'permanent odour', not appreciating that this is the shortest way to permanently

57

damage their sense of smell. The coarse formulations of many of the commercial products—such as a combination of di-para-di-chloro-benzene with a floral scent—so fatigue people's noses exposed to the compound smell that they practically lose their olfactory perception over a fairly wide range. Their faculty of appreciation and enjoyment of odours and perfumes becomes blunted since they are engulfed by smells incessantly. Their perception of odour recedes into the background and no longer causes conscious discrimination. Artificial odours have become a subconscious part of the environment, and are as indispensable as the radio noise labelled music without which social life—and often just life— is intolerable for many.

Man's preoccupation with smell has taken on a new turn with re-odorization. Not satisfied with the natural aromas and scents, the flavour chemist can impart any particular odour—or taste—to the simplest object like a pair of stockings. As an experiment, one and the same type of stockings were divided by the manufacturer into two lots. The one was left its natural odour of nylon which is almost imperceptible in any case, the other was completely deodorized and then re-odorized to smell like a genuine silk stocking. Both lots had the identical label which read 'Smell its quality'. The 'silk' stockings were sold out in the shops within hours whereas the others, at the same price, were left in their trays.

The implications are obvious. Subliminal vision in the cinema, subtle re-odorization of goods, colourful packaging—it all adds up to 'sales psychology'.

Smell can also be a social problem in a different way. One instances the person who is hypersensitive to a certain smell, such as faeces, which would prevent her from using the toilet in her house for at least an hour after it had been used by another person (108), or the woman, a physician herself, who could tell her husband with infallible accuracy where he had been simply by the odour retained in the texture of his clothes, or on his skin (109). The implications of odours in human behaviour are only beginning to be studied in a coherent manner and this new branch of investigations is likely to produce some most interesting results (106).

8 Pure air and stale air

In contrast to polluted air, there are 'clean air' and 'pure air'. *Clean air* is a modern technological term describing the air in a Clean Room which is required for such operations as the production of transistors, the assembly of space gear, and others. *Pure air* is best defined as air free from any matter save that recognized as a standard ingredient of pure air (Table 15). Of these components all but three are chemical elements. Two of the others, carbon dioxide and methane, are of biological origin, while ozone forms from atomic oxygen, which occurs not below altitudes of 100 km (60 miles), through the action of ultra-violet radiation incident from the sun. Being heavier than air (1·658), ozone sinks to a level of about 30–50 km (20–30 miles) above ground where it floats on the denser atmosphere below.

Carbon dioxide is the respiratory product of animals. In pure air, industrially produced carbon dioxide is disregarded.

Methane is present in the troposphere, the atmosphere immediately above ground up to 10 km (6 miles) high at the poles, and 17 km (11 miles) at the equator, at a volume concentration of about 1·5 ppm. Total production rate is estimated at $2·7 \times 10^{14}$ g methane per annum (312). This is about 75% from biological, and 25% from industrial sources. LIBBY, and later others, have measured the ^{14}C content in methane showing that it is derived as 75% from recent wood. Measurements of methane collected by high altitude aircraft show a considerable variability with altitude and time (208).

In contrast to the above attempt at specifying what might be labelled 'natural pure air', there is a 'natural fresh air' standard proposed by TURK and D'ANGIO (44) in a paper to the American Air Pollution Control Association at their 54th Annual Meeting at New York, in 1961. They describe mountain air as being very pure, and make this purity the cause of the specific organoleptic effect of fresh mountain air rather than suggest

—as others do—that it is the presence of aromatic substances which give mountain air its character. It is descriptions like this which make it almost impossible to work out a specification of articial fresh air. With the annual growth rate of aerosols increasing by more than ten million particles per cm^2 even in Davos, Switzerland, nearly 1600 m above sea-level, it is questionable whether mountain air can really be considered pure.

Apart from gases in trace, and some in larger quantities, the composition of standard pure air is:

By weight $76.8\% \, N_2$ $23.2\% \, O_2$
By volume $79.1\% \, N_2$ $20.9\% \, O_2$

The mean molecular weight $M = 28.84$; standard density $= 1.00000$; specific gravity $= 1.293$ g/l $= 1.293$ kg/m^3 $= 0.081$ lb/ft^3 at 0°C and 760 mm Hg ($= 1$ atm $= 29.92$ in Hg $= 14.7$ lb/in^2).

Table 15   *Composition of pure, dry atmospheric air (32, 33)*

Component	% by weight	% by volume
Nitrogen	75·54	78·084
Oxygen	23.14	20·946
Argon	1·27	0·934
Carbon dioxide	0·05	0·033
Neon	0·001 2	0·001 818
Helium	0·000 7	0·000 524
Krypton	0·003	0·000 114
Hydrogen	0·000 04	0·000 05
Xenon	0·000 036	0·000 008 7
Ozone	0·000 001 7	0·000 05
Methane	0·000 083 1	0·000 15

The admixtures usually found over residential and industrial areas change slightly in quantity with relation to seasons and local conditions, but generally consist of water vapour, carbon monoxide, sulphur dioxide, hydrogen sulphide, ammonia, nitrates, aerosols, mists, and a variety of gaseous effluents from trade waste to which must be added micro-organisms and organic matter such as respiratory products of vegetation apart from CO_2.

These last are volatile substances, strongly aromatic, and belong to the terpenes of general composition $C_{10}H_{16}$. They are present in the atmosphere in concentrations from 2–20 ppb (20×10^{-9}) and are responsible for the blue haze that forms over mountains. A total of about 1000 million tons of terpenes are released per year by vegetation over the whole world. It has been suggested (308) that the dissipation of terpenes and other volatile plant products pass through the same cycle as the dissipation of petrol vapours in cities. The blue appearance of the naturally occurring haze is due to TYNDALL scattering of blue light (short wavelengths) by submicroscopic particles. Sunlight alters the terpenes photochemically forming ozonides

$$=C=C= \; + \; O_3 \; \rightarrow \; =C\overset{\displaystyle }{\underset{\displaystyle O_3}{\diagdown \diagup}}C=$$

and peroxides. Free radicals are also produced by the sun's effect and lead, later, to the formation of AITKEN (condensation) nuclei. The terpenes can be measured quantitatively by a gas chromatograph, and their photochemical transformation into particulate matter by a nucleus counter. SMITH (310) finds some evidence for the toxicity of terpenes.

A considerable number and variety of gaseous reactive hydrocarbons contained at low concentration in air, are converted by short time irradiation with solar ultra-violet into the aerocolloidal state. Only short exposures are required for this transformation (311).

The vertical distribution of atmospheric components has been investigated by many workers. Ozone concentrations on mountain tops and in valleys vary by a factor of 3–4 (104). Measurements of aerosol distribution in a coastal area (Cape Blanco, USA) indicate that aerosol formed from sea spray does not penetrate to altitudes above 2000 m. There are also considerable concentrations of sulphur, chlorine, and other substances (105).

Standard breathing air is assumed to consist of nitrogen, oxygen, carbon dioxide, and argon as to over 99·99 %. The other gases occur as natural admixtures to which must be added such pollutants as there are. Their concentrations in air should not exceed maximum acceptable concentrations (MAC) for each constituent.

It is of interest to compare with that simple statement the composition of standard drinking water as compiled by the World Health Organization, Table 16.

Phenolic substances are expressed as phenol. Carbon chloroform extract describes organic pollutants. Alkyl benzyl sulphonates describe surfactants (detergents). Colour units are on the platinum–cobalt scale. Turbidity is measured in turbidity units.

Table 16   *International standards for drinking water (313)*

Component	Maximum acceptable Concentration mg/l	Maximum allowable Concentration mg/l
Total solids	500	1500
Iron	0·3	1·0
Manganese	0·1	0·5
Copper	1·0	1·5
Zinc	5·0	15
Calcium	75	200
Magnesium	50	150
Sulphate	200	400
Chloride	200	600
Mg + Na sulphate	500	1000
Phenolic substances	0·001	0·002
Carbon chloroform extract	0·2	0·5
Alkyl benzyl sulphonates	0·5	1·0
Colour	5 units	50 units
Turbidity	5 units	25 units
Taste	unobjectionable	—
Odour	unobjectionable	—
pH range	7·00–8·5	not less than 6·5 or greater than 9·2

The chemical composition of the soil and a rural environment pollute water collected for drinking purposes. Often the colour is a reddish brown from dissolved iron salts, and micro-organisms from animal droppings may be found in quantity though those bacteria, mostly coliforms, do not always constitute a danger to health. However, there may also be some salmonella which will cause enteritis or, at least, digestive upsets.

The volume of air inhaled per 24 h day is 20 m³, of which one half is required during the 8 working hours, and the rest during the remaining 16 hours. Of pure air, oxygen is the active component and any reduction in its availability makes itself felt immediately. The presence of 'foreign' matter in air, including carbon dioxide, will cause such a reduced intake of oxygen. However, since CO_2 is a naturally occurring constituent of air, man's respiratory system is tuned to its intake. In general terms, breathing takes in oxygen and releases carbon dioxide, so that any given volume of air in which an animal or a person breathes becomes slowly unfit for the purpose.

Table 17 *Composition of respiratory air*

	Volume % of	
	Inhaled air	Exhaled air
Oxygen	20·95	16·4
Nitrogen	79·01	79·5
Carbon dioxide	0·04	4·1
	100·00	100·0

During a day's breathing a person requires

oxygen	4·19 m^3
nitrogen	15·802
carbon dioxide	0·008
	20·000 m^3

and exhales

oxygen	3·28 m^3
nitrogen	15·90
carbon dioxide	0·82
	20·00 m^3

The maximum acceptable concentration (MAC) or recommended threshold limit value of carbon dioxide is 5000 ppm = 0·5% v/v. The carbon dioxide content of fresh air (0·04% v/v) is well below the limit, but the CO_2 content of exhaled air (4·1% v/v) is higher than the MAC by a factor of eight. Exhaled air is a proper asphyxiant.

The balance sheet of respiratory air shows a loss of oxygen.

Oxygen	+	3·28 −	4·19 =	−0·91 m^3
Nitrogen	+	15·90 −	15·802 =	+0·098 m^3
Carbon dioxide	+	0·82 −	0·008 =	+0·812 m^3
		(exhaled)	(inhaled)	+(surplus)
				−(loss)

The loss of oxygen from the total air volume (20 m^3) is made good—metaphorically—by the gain in carbon dioxide and nitrogen. Life of the type we know could not be maintained without the combined effects of plant ($CO_2 \rightarrow O_2$) and animal ($O_2 \rightarrow CO_2$) respiration. Loss of oxygen from respiratory air is also a cause of loss of odour perception. A reduction

by 20 % of the oxygen in every 20 m³ of air produces senescence phenomena in man; loss of judgment, sensory capacity, insight and memory.
These failures occur regardless of age, sex, and race. They are caused by
oxygen starvation of nervous tissue. Unless a room is well supplied with
fresh air, people's breathing will soon reduce the oxygen content to below
normal value with the consequent loss of capacity to smell.

An occupied room becomes 'stuffy' when the amount of carbon dioxide
in it reaches $0 \cdot 1 \%$. For good ventilation each person should be allotted
25-30 m³ of breathing space and should be supplied every hour with
60-80 m³ of fresh air. Ventilation rates vary with occupation. If the CO_2
content is to remain below $0 \cdot 1 \%$ the following ventilation rates are
recommended:

Occupation	Ventilating rate
At rest	50 m³/h
Moderate work	80
Strenuous work	160
(in round figures: 1 cfm = 17 m³/h)	

The empirical formula for deriving the maximum rise of carbon dioxide
in a room after long hours of occupation by P persons is $70P/R$ where
R is the rate of ventilation expressed in ft³/h, or $2P/R$ when measured
in m³/h.

In contrast to pure, or to fresh, air is stale air which is a characteristic
of populated—and overpopulated—areas, be they residential, commercial, or industrial. While the word 'stale' carries little technical information, the term 'used-up' air is self-explanatory. It is air which has been
repeatedly in- and exhaled by people. Stale air is deprived of its full
oxygen value and carries a considerable surplus of carbon dioxide. It is
also laden with vapours of various kinds (sweat, alcohol, urine) and may
also carry ultrafine particulate matter, such as tobacco smoke. Cos-
metics, cooking odours, metabolic odours, and others contribute to its
characteristics. People working, or living, in stale air are oblivious of its
presence. Their sense of smell is fatigued so that they have no appreciation
of that particular brand of air pollution, but this fatigue is restricted to
the smell in their habitual surroundings. They react quite normally to
other odours.

Perspiration is the most common of metabolic sources of odours. A
person working in a hot and humid environment will lose up to 900 g

sweat per hour. Sweat has a salty-sour taste. Normal sweat from healthy persons does not smell of any particular odour, but it may become repulsive if the urea content, normally about 0·03%, or butyric acid, normally about 0·001%, increase but slightly. Sweat consists of 99·5% of water and has sufficient salt (sodium chloride) to lose 3–4 g of chlorine per hour in profuse sweating. Sometimes valeric acid too is found in sweat and this, together with butyric acid, forms the characteristic body odour. Normal, non-smelling perspiration is produced by the sudorific glands of which there are two million in the skin of an adult person. There are other skin elements which secrete an offensively smelling fatty substance. The apocrine cells are situated in the groin, around the anus, around the lactating mammae, and in the axillae.

There are pathological conditions which are characterized by offensively smelling sweat, such as in bromidrosis which is produced by decomposing keratin of the feet. The disorder is due to the invasion of the skin by staphylococci, and the smell develops due to alkaline perspiration. The genital region and axillae may, too, be sources of bromidrosis. Whilst this condition is purely of physiological origin, excessive sweating, though free from offensive odours, may be brought on by fear, worry, and other conditions of mental disturbance. This hyperidrosis can occur after the intake of spiced food, or in excessively heated rooms, but also at low temperature. Further bad ventilation is itself a highly effective source of hyperidrosis. A variety of diseases may contribute to the pollution of air by smells: cancer in advanced stages, osteomyelitis, tinea versicolor, ozaena, and others. Disturbance in the urinary system, for instance cystitis, result in ammonia being excreted in fresh urine, a fact to which the human sense of smell reacts immediately. A malfunction of protein metabolism causes alkaptonuria, and diabetes will produce acetone in the body which is excreted in urine. Stale, or fermented, urine smells not only of ammonia, but also contains cyclohexen-3-one (urinoid) which is highly objectionable.

Foods also impart strong characteristic odours to faeces, metabolic gases, and to urine. The food itself may be nearly odourless, or have an odour totally different from the metabolic smell it causes. After having eaten asparagus, the urine will strongly smell of methyl mercaptan.

Personal deodorants to kill the smell of sweat or more intimate odours are a questionable relief, if relief they are at all. Frequent use of water and soap may be advisable, but is not the general panacea some people make it out to be. For instance, abnormal vaginal secretion may need to be treated as a disorder, not as a matter of defective hygiene, or a case for the application of deodorants.

Odours from the alimentary tract depend on the type of digestion taking place. When stools are alkaline they indicate protein putrefaction and give off highly offensive smells. Bacterial, as opposed to fermentative, decomposition produces skatole and indole, both faecal smells, but different from one another. Intestinal gas contains hydrogen sulphide, skatole and indole; mercaptans may also be present. Their composition is determined by the type of food and intestinal flora (bacteria); the quantity depends, largely on the amount of meat eaten.

Stale air is not only characterized by a low oxygen and high carbon dioxide content, by a certain proportion of metabolic waste, but also by cooking odours. The objection to cooking odours, whether from vegetable or animal matter, stems from the production of sulphur compounds which then pervade the atmosphere together with a considerable quantity of steam. On condensing on cold walls or fabrics the droplets are fairly charged with sulphurous substances and are absorbed into the fabric of furniture, carpets, wallpaper, and similar porous materials. The condensed droplets will soon evaporate when suitable conditions prevail, leaving behind a layer of osmogenic molecules. Such layers are also produced even in the absence of steam as a carrier. In dry heat, airborne molecules may be dissolved in the imperceptible layer of water which covers all surfaces in temperate climates. This atmospheric moisture is similar to the layer of imperceptible perspiration which covers the entire skin surface. It has been found that the retentivity of a surface for adsorbed or absorbed molecules varies with the material of the surface and the type of molecule. Skatole is retained on metallic surfaces up to three weeks. Glass has but little retentivity for any molecule under atmospheric conditions.

Meat contains and, when boiling, gives off, dimethyl sulphide, $(CH_3)_2S$. A more complex molecule, dimethyl disulphide, $(CH_3)_2S_2$, is produced by boiling cabbage. In the fresh raw state, cabbage contains the thermo-labile free amino acid L-S-methyl cysteine sulphoxide. Upon being heated, it breaks down forming the offensively smelling sulphide. Proteins, too, produce strongly smelling decomposition compounds. Protein consists of long-chain amino acids with an average content of 0.8% sulphur. These breakdown products give venison, cheese, and other foods, their high flavour. Acetylmethyl carbinol, $CH_3.CO.CHOH.CH_3$, is a typical cheese flavour, but when oxidized to diacetyl, $CH_3.CO.CO.CH_3$, it gives fresh butter its attractive odour and taste. Acetylmethyl carbinol is contained in Gorgonzola, very ripe Stilton cheese, or similar types, at concentrations up to 8–9 mg/100 g. In very mild cheese, its proportion is only 0.1 mg/100 g. Milk and cheese contain β-methylmercaptoproprionalde-hyde. In concentration, the compound smells of raw pumpkin, in dilution like cheese. The slightly cheesy flavour in milk which has been standing in the sun is also caused by this compound.

Fish smell of amines, and eggs of sulphides. Both are objectionable in concentration. The fumes from hot, and more so from overheated, cooking oil have an offensive fatty smell of the acrolein type. Oils have a strong absorptivity for other substances, and then become tainted by the odour of the absorbed matter. Frying fish in oil without batter will produce a heavy fatty smell which has a strong overlying note of amine.

The eating of highly spiced and flavoured foods is often characteristic of whole ethnic groups of people. The eating of raw onions and garlic taints the breath in a manner which is quite unacceptable to many people. Again, it is disulphides which produce the smell, i.e., diallyl sulphide, $C_3H_5.S.S.C_3H_5$, and allyl propyl disulphide, $C_3H_5.S.S.C_3H_7$.

Roasted coffee beans emit molecules of

Acetaldehyde	$CH_3.CHO$
Diacetyl	$CH_3.CO.CO.CH_3$
Furan	$CH.CH.CH.CH.O$
Furfuraldehyde	$C_4H_3O.CHO$
Furfuryl alcohol	$C_5H_5O.CH_2OH$
Furfuryl mercaptan	$C_5H_5O.S.CH_3$
Hydrogen sulphide	$H.SH$
Methyl ethyl carbinol	$CH_3.CH_2OH.CH_2OH.C_2H_5$
Pyridine	$CH.CH.CH.CH.CH.N$

in various proportions which are responsible for the aromas of the various blends. Most of these molecules are highly offensive osmogens, such as hydrogen sulphide, and furfuryl mercaptan. Caffeine, $C_8H_{10}O_2N_4$, upon being overheated during the roasting process, breaks down and one of the fragments will combine with hydrogen—from steam—to form acrolein (allyl aldehyde), $CH_2.CH.CHO$, which has a powerful pungent odour.

The presence in foods of such strong and objectionable odours as listed above is not a singularity. On the contrary, where these are absent by nature, the flavour chemist will add certain strongly, even offensively, smelling compounds to give food a 'kick'. Chocolate, without a minute trace of indole (a strong faecal smell) would taste flat and uninteresting.

Stale air is especially offensive at normal ambient temperatures. At low temperature, not only is olfaction less sensitive, but also is the output of the various sources contributing to stale air reduced. Man is not only a source of odours, metabolic and artificial, but also of heat.

Table 18   *Emission of thermal energy from the human body*

Body surface	74·0 %
Expired air	2·5
Faeces and urine	1·5
Insensible perspiration	12
Pulmonary perspiration	10
Total intake = total output	100·0 %

Heat is given off by convection in that the air surrounding the body cools its surface; also by radiation in that the body at surface temperature 37°C radiates thermal energy into its environment which is, usually, at a lower temperature. Finally, heat is lost by evaporation from the skin. Small amounts, about 2 % or less of the heat input, are used for heating up ingested food, and pulmonary air. Less than 1 % of heat is lost in urine and faeces.

Table 19   *Heat losses of the human body*

Loss by	%
Convection	31·0
Radiation	43·74
Evaporation	21·71
Heating ingested food	1·55
Heating pulmonary food	1·3
Loss in urine and faeces	0·7
	100·00

Applying WIEN's Displacement law $T\lambda = 2898$, the wavelength of emission of the human body (average skin temperature 33°C) is centred at $2898/306 = 9\cdot44$ μm, i.e., in the infra-red region of the spectrum ($T°K = 273+33 = 306°K$). It is generally accepted that healthy individuals radiate in the overall range 2–12 μm with a peak at about 9·5 μm. On closer investigation it is found that there are two narrower regions of emission, 2–5 μm, and 7–11 μm. Below 2 μm, and between 5–7 μm there is hardly any radiation since the water of the body absorbs strongly in these regions of non-emission. The skin, irrespective of colour, and with regard to the longer waves, behaves as a perfect Black Body. It radiates at a rate proportional to the fourth power of its own temperature (STEFAN's law). The energy emitted is calculated as

$$E = \delta(T^4 - T_0^4)$$

where T is the absolute temperature, $^\circ K$, of the body (average $306^\circ K$),
T_0 is the absolute temperature of the environment ($15^\circ C = 288^\circ K$),
$\delta = 5 \cdot 67 \times 10^{-8}$ $W/m^2/^\circ K^4$ (STEFAN-BOLTZMANN constant expressed in Système International (SI) units).
Thus, the energy emitted $E = 5 \cdot 67 \times 10^{-8}(307^4 - 288^4) \approx 107$ W/m^2.

Average total body surface is $1 \cdot 8$ m^2, assuming height of person 170 cm, weight 70 kg (after the formula of DU BOIS and DU BOIS) (41). But only the area of projection, i.e., about 70–80 % of the total surface, radiates. Total emission ranges from about 150–200 W. This value varies within the limits of weight and height by $\pm 10\%$. From the variable quantity in STEFAN's law it is apparent that with increasing ambient temperature T_0, the emission of heat from the body will decrease. The person begins to perspire severely the smaller the difference $T^4 - T_0^4$. At uncomfortably high temperatures, theoretically above skin temperature, the body ceases to emit heat altogether, and will now absorb heat from the environment, making things worse. The values above apply to the bare body. Taking clothing into consideration with its great variety of materials (animal, vegetable, artificial fibres all having different thermal coefficients), and not forgetting the dependence of heat generation on mental stress, it will be appreciated that such values are valid approximately only unless exact conditions of test are specified.

The range of heat transfer from the average person to his standard environment ($20^\circ C$, 760 mmHg) can be assessed from the figures of Table 20.

Table 20   *Occupation and heat transfer*

Resting	100– 110 kcal/h
Writing at a desk	125– 130
Typewriting	140– 145
Standing upright, or doing light work	150– 200
Working hard (machine tool operator, assembly work)	250– 300
Taking strenuous exercise	1000–1500
[1 kcal = 4·184 kJ; 1 kcal/h = 1·16 W] (42)	

For purposes of comfort engineering it may be assumed that the human body gains heat from his environment at ambient temperatures above 25°C (78°F), and loses heat to its environment at ambient temperatures below 18°C (62°F). Heat losses also vary with posture; crouched together the body radiates a minimum of heat, and losses are smallest; stretching increases the radiating surface.

Figures of heat loss to the environment are of special significance when considered in conjunction with room temperature, and how they affect working efficiency.

Table 21   *Effect of heat loss*

Room Temperature °C	Physiological effect on healthy adult
17–22	Comfortable in winter
18–24	Comfortable in summer
10	Physical stiffness of extremities
18	Optimum conditions
24	Physical fatigue begins
30	Mental activities slow down, errors are made
50	Mental and physical incapacity sets in; conditions are tolerable for 1 hour
70	No activity of any kind possible; conditions are tolerable for 30 min. These temperatures apply within a humidity range of 30–70% R.H.

Working at high ambient temperatures in humid air will cause a person to produce about 900 g of sweat per hour. For the evaporation of this quantity of sweat 530 kcal/h = 615 W are required.

All the above data are compiled for pure, i.e., unpolluted air, unless stated otherwise. If the air carries heat absorbing matter, whether in molecular or particulate form, temperature will rise steeply.

Table 22   *Heat absorption of polluted air*

Air, pure	Heat absorption =	1
Air, polluted with patchouli		32
Air, polluted with oil of roses		36
Air, polluted with oil of anis		372

The more stifling the air, the greater is its heat absorption.

9 Weather

It is unquestionable that weather affects the distribution of airborne pollutants. Temperature, inversions, sultry days, summer haze covering large areas, and similar weather conditions, but also rain, wind, and thunderstorms control to a great extent the character of local air pollution.

The times when the effects of weather were accepted like other 'acts of God' seem to be coming to an end. At least, every effort is being put into the plans to modify and control the weather to suit particular requirements of man. A Bill was put before the United States House of Representatives at the 90th Congress (1st Session, House of Representatives, 26th April, 1967) to authorize the Secretary of Commerce to carry out a comprehensive programme in the field of weather modification as relates to the control of air pollution 'and other similar deleterious aspects of urbanisation upon the composition of the atmosphere' (324). At the AAAS Symposium on Weather Modification which was held in New York on 30th December, 1967, a paper was read outlining the principles of weather modification, and the implications of the Bill were discussed (323).

10 Vegetation and air pollution

Although osmogenic pollution of the air is not a particular hazard to vegetation, air pollution of the particulate type is. Smoke and sulphurous or other fumes are injurious to plant health as many a hedgerow along a motor highway can demonstrate. The biological aspects of certain chemical wastes are interesting in that they act as phytotoxins, but are no health hazard to man. The antiknock compound tetraethyl lead (lead alkyl, $Pb(C_2H_5)_4$) is a normal constituent of the hedgerows along motor ways. This applies in particular to the highways in the United States because of the density of cross-country motor traffic. At major road junctions ashed grass was found to contain 3000 ppm of tetraethyl lead, and up to 150 yards downwind the concentration was still 50 ppm (240). Vegetables grown on a farm within a distance of some 10 m from a busy road contained an average of 100 ppm of the antiknock compound.

A simple quantitative method allows the determination of between 0·4–0·8 μg of lead per m^3 of air. A quantity of 2 parts tetraethyl lead from air is absorbed in concentrated sulphuric acid which decomposes the compound. The first two of a series of absorbers are cooled to about 0°C. The lead is detected with dithizone, $C_6H_5N:N.CS.NH.NHC_6H_5$, which dissolves in sulphuric acid with a blue colour. On reacting with lead this turns brick red (204).

In a recent literature survey of the health aspects of airborne lead emissions (241, 256) no conclusive evidence of deleterious effects in man was found. However, concern was expressed rather strongly about long-term effects and the possibility of chronic disorders though they may remain on a subclinical level for a considerable time. An increase in lead is viewed with some apprehension.

The leaves of roadside vegetation, are covered in dust and can breathe only with great difficulty, even if the rain washes off a little of the dirt. The soil, too, is covered with the dust of the road and, being porous, acts as a hugely efficient adsorbing medium for the exhaust gases from petrol and diesel engines. Rain dissolves the matter which is then washed down into, and adsorbed by, the soil, poisoning the earth and the roots of plants.

This is of interest to the town and country planner, and to the engineer who designs sewage works, chemical plant, and other industrial premises which, for local reasons, are built in the country and must be hidden from sight by screens of trees. It is therefore of importance to plant only those trees in industrial areas which stand up well to air pollution. Deciduous trees can cope better with polluted air because they shed their leaves for 5-6 months a year and are exposed to respiratory hazards for only half their lifetime, unlike conifers which keep their needles for several years. A summary of how trees and vegetation react to air pollution is given in a WHO publication (233).

Vegetation also absorbs sulphur. Examination of the needles of conifers along roads in Yokohama showed that their sulphur content was almost proportional to aerial pollution with sulphur dioxide. In the polluted zone, the sulphur content was from 2-4 times greater than in the non-polluted zone, namely for: *Chamaecyparis pisifera* (feathered cypress) 0·50%, *C. obtusa* (miniature feathered cypress) 0·39%, and for *Cryptomeria japonica* (Japanese cedar) 0·43% (255).

The following stand up to smoke in industrial areas:

Deciduous trees:
Ash, Beech, Elm, Gean (Dwarf Cherry), Holly, Locust or False Acacia, Field Maple, Mountain Ash, Oak, Plane, Poplars, Sycamore, Walnut.

Coniferous trees:
Cypress, Larch, Firs, Scots Pine, Yew.

Part II

LEGAL ASPECTS AND EVALUATION

11 Legal aspects

From earliest times, human occupations have been subject to legislative attention. Regulations in the cities of Harrapa and Mohenjo-daro in the Indus valley, about 2500 BC, controlled the setting up of manufactures in quarters of the town to windward of residential districts so that neither smoke, noise, nor smell would descend upon the temples and dwellings.

Throughout the centuries legislation and the social life of the population affected one another deeply. It always was a principle of the lawgiver to couch his provisions in general terms, leaving it to the judge to interpret as circumstances required.

In 1956, England made legal history by introducing technological specifications and instructions into the language of the legislator. The Clean Air Act (4) which came into force on 5th July, 1958, explained how to measure, objectively, smoke density, and introduced the Ringelmann Chart as a smoke density standard. Consequently, this instrument was also admitted as a proof in Courts of Law. Following upon this unprecedented step, the Noise Abatement Act (5) came into force on 27th November, 1960. There, again, the evidence of a technical instrument, the sound level meter, was made acceptable in Court.

The Clean Air Act of 1956 has subsequently been extended in both legal and technical matters, but no mention of odour has been incorporated in any of the later legal instruments. The Law of Nuisance is still the only one applicable to offences by odour. Though various aspects of air pollution by smoke have a bearing on the question of air pollution by smells, any legislation made is primarily concerned with the optical nuisance caused by smoke, while its acrid smell is considered purely incidental and of no interest to the legislator. Smokeless fuel may cause little or no smoke and soot, but nobody bothers whether it might be an odourless fuel at the same time—or whether it might be possible to produce such a fuel at all.

The entire Clean Air Act 1968 became operative with effect from 1st October, 1969. Guidance and advice on the application of the various provisions are provided in Circulars 54/69 (for England), 51/69 (for Wales), and in SDD Circular 49/69 (for Scotland). Statutory Instruments No. 995 (C.24) (England and Wales; Ministry of Housing and Local Government) and No. 1006 (C.25) (Department for Scotland) apply (HMSO 1969).

The Secretary of State for Local Government and Regional Planning has set up a permanent central unit for coordinating activities controlling environmental pollution. A permanent Royal Commission has also been established whose terms of reference are 'to advise on matters, both national and international, concerning the pollution of the environment; on the adequacy of research in this field; and on future possibilities of danger to the environment'. The Royal Commission can take up any problem relating to pollution so that the best advice will be available to Government Departments responsible for executive action.

Whereas in English Common Law a nuisance must be committed, and be confirmed as such in Court before proceedings can be taken against the perpetrator, Statute Law—such as the Clean Air Act or the Noise Abatement Act—make things much easier. If smoke density in a given case is above the permitted limit which is clearly and unambiguously expressed as being Ringelmann shade x, then no argument is necessary or possible. The reason for exceeding the statutory limit may have extenuating circumstances, or not, but *de facto* and *de jure* there is no argument.

There is no legislation specifically against odour offences on the English Statute Book, nor is there likely to be one in the near future. This is not the fault of the legislation, but of science. Nobody has been able, as yet, to produce an instrument for the indication of smells, their measurement, and their evaluation. There is not even agreement with regard to specifications of odours, what exactly constitutes an obnoxious odour, or how to define its intensity unambiguously and in terms permitting of repetition. In the absence of an 'artificial nose' or, more specific, an 'electric nose', there would be little purpose served if the present practice of defining a nuisance were to be replaced by some new definition which could improve neither clarity of expression nor scope of the offence. Although there are variously worded definitions, and comments on definitions of a nuisance, available in legal literature, the most explicit one is that quoted by Halsbury, in Laws of England, vol. 28. This is known as the Rule of Rylands v. Fletcher, 1868, and defines that a nuisance is committed if 'the thing which does the damage has escaped'.

Although originally applied to arguments involving an underground water reservoir which had burst open and flooded some of Ryland's property, the definition is so ingenious that it is generally applicable and still serves its purpose.

The Common Law of England is based on centuries of precedence and is adequate in all those cases which, in principle, differ but little from the ways of English life in times past. But it is often inadequate where technical or scientific facts in modern connections are involved. This is simply because there is no precedent on which to base jurisdiction. Statute Law is the modern supplement to Common Law.

In England, the Public Health Act, 1936, empowers the Local Authority to take action against offenders, and serve an Abatement Notice on the person or Company responsible for a nuisance. In the case of non-compliance, the Authority can then proceed by laying a complaint before a Magistrate. If the nuisance is proved in Court, a Nuisance Order is made. Allowing the nuisance to persist or recur then becomes a punishable offence.

A long time before the Public Health Act, the Alkali Works Act, 1863, gave powers for controlling obnoxious odours, in fact noxious emissions of all kinds from industrial plants, mainly chemical works; and the central inspectorate was set up soon after. It was the Public Health Act of 1875 which specifically dealt with excessive smoke emission from industrial premises. The word 'excessive' ought to be noticed because, in 1875, and for more than half a century afterwards, there was no definition of excessive smoke as there was no Ringelmann Chart, or equivalent, although the public never lacked in interest and tried repeatedly to propose relief. In 1882, Dr. Edward Ballard published a report about 'the effluvium nuisances which arise in connection with various manufacturing and other branches of industry, especially with regard . . . to the degree in which the nuisance can be arrested', but, as in many more instances, proposals were either impracticable, inapplicable, or not suited in other ways to bring real relief from the nuisance, or did not even try to offer a specification on which to formulate an Act of Parliament. As the years passed, the Public Health Act was steadily amended and its scope widened. The Act was successfully used in the fight against dense smoke emission, and in spite of the inherent limitations, it gave guidance to industrialists.

There is a remarkable proviso built into the Act of 1936. If a Local Authority should be unwilling to take to Court an offender who has not complied with an Abatement Notice, a request signed by a minimum of ten local residents affected by that decision can force the hand of the

Authority. Even if the latter have been successful in winning the case, the premises of the offender will never on that account alone be closed down although, in law, the Authority could do this. The offender has a right to demand compensation for any disturbance, and since public funds cannot be so employed recourse is usually taken to offering the defendant new premises well away from residential areas. If he accepts, all is well. If he refuses, the Local Authority has now a good case to get the premises closed down by a court order without incurring the risk of being sued for compensation.

In a court case dealing with complaints about offensive odours people will be called as witnesses. It happens many a time that a witness, intending to strengthen the case of the plaintiff, will vouch for the disagreeability of the sensation by emphasizing that he felt quite ill every time the dreadful smell was perceived.

Not only will medical evidence claim that sickness is not caused by smells unless they are toxic or accompanied by pathogenic bacteria, but legal precedence will refer to the dictum of Lord M. R. Romilly in Crumpet v. Lambert (1867, 15 L.T. Rep. 600, 601; L.R. 3 Eq. 409, 412) that injury to health is not a necessary ingredient in the cause of action for nuisance by smell.

A nuisance is sometimes described in legal textbooks as 'a disturbance of another man's right' without defining that to breathe clean non-polluted air is man's right. However, this was the opinion upheld by J. Veale in the case of Halsey v. Esso Petroleum Company Ltd (1961, 2 All E.R. 145) and his dictum has now included as a nuisance any action which, ultimately, will deprive a person of this right.

The defendant may take the line of least resistance and claim that his premises, or trade carried on on leased premises, do not produce more, or worse, smells than somebody else in the immediate vicinity; and this other person has not been prosecuted; so why should he be? Alas, his efforts are bound to fail because of the dictum of Lord Romilly, but also because Vice-Chancellor Knight-Bruce said that a nuisance is the thing which 'materially interferes with the ordinary physical comfort of human existence—according to plain and sober and simple notions obtaining among the English people' (Walter v. Selfe, 1851, 4 de G, and Sm. 315, per Knight-Bruce, V. C.).

The battle against air pollution is fought on all fronts, including the legal one. The difficulties facing legislation are like those confronting the engineer. There are changes and improvements, and the law, in order to remain effective over a long period of years, ought to anticipate them all. This is impossible and one is inclined to take the view that air pollution laws, as they are at present, are mere stopgaps to gain time (110). One cannot do more than study the trend of legislation in the hope that in years to come there might be more facts and data available which will be suitable for incorporation in legislation, and that instruments, and an internationally recognized and unified theory and terminology for the measurement of odours will be available.

The Manufacturing Chemists' Association, Washington, issues from time to time a collection of notes and comments called *Source Materials for Air Pollution Control Laws* (111). In 1965, for instance, some 400 bills on air pollution control were introduced. *Source Materials* includes the full text of six current state statutes (California, Illinois, Louisiana, Texas, Pennsylvania, New York) which illustrate different approaches to state legislation against air pollution. Of these, Illinois, Louisiana, New York, and Texas exercise state control, California regional control, and Pennsylvania combines both types of control. Common to all state legislation is the principle that public health must under no circumstances be affected by a given case of air pollution. Medical and scientific evidence and consensus of opinion is the touchstone.

If public health is not affected, the general benefits to the community must be balanced against the cost of the proposed improvement. Ultimately, this cost is borne by the community, whether it knows it or not. The degree of abatement is decided upon by the Authority with the assistance of an expert panel, and will influence the cost of such abatement. The findings of the Local Authority can be fought in a court of law. The Attorney General is authorized to apply for an injunction against a person found guilty of causing a violation in order to prevent him from continuing, or committing any further violation of the regulations. Persons other than the State have no rights, and can not acquire any, by virtue of the respective Act. Any regulations or other determinations of the local Board do not create any presumption of fact or law.

Regional and local air pollution control may be permitted in place of, or in addition to, state control, but in all cases regional or local jurisdiction is required to be consistent with the statute. State control is always dominant.

After detailing these statements for each of the six States, excerpts of 6-10 pp. each from the respective Acts are quoted.

The constitution of the various Control Boards is given in every instance and part of Section (4) of the Illinois Air Pollution Control Act is quoted here giving specific instructions:

'The members of the Board shall be: (1) The Director of the Illinois Department of Public Health, (2) a registered professional engineer with at least five years' experience in the field of air pollution control, (3) a physician experienced in the field of industrial medicine, (4) one engaged in a field directly related to agriculture or conservation, (5) one engaged in the management of a private manufacturing concern, (6) one representing labour, (7) one engaged in municipal government, (8) two selected at large.'

More and more of the United States prepare laws against air pollution. Among the most recent, the Air Quality Control Regulations have been enacted by the State of Maryland (28th January, 1969). The salient features of this, as of all similar regulations by other States, are the requirement for all significant sources of air pollution to be registered with the Department of Health, and to observe pollution limits. The emission of obnoxiously smelling vapours and fumes is mentioned in particular.

Probably the most important step taken by the Federal Government of the States is that part of the Anti-Air pollution legislation which concerns Interstate air pollution, caused by sources which are near common borders. Recommendations of the Department of Health, Education, and Welfare for its control, issued on 14th March, 1969, by the National Air Pollution Control Administration, are based on proposals issued by the Air Pollution Abatement Conference held in the summer of 1969 at Ironton. If remedial action on the lines of these recommendations is not taken, the Attorney General may initiate action in United States courts.

The gist of interstate legislation on a federal basis is, briefly, the dictum that the several States having a common border must cooperate in reducing both emission and immission of the nuisance and maintain the conditions set out as a standard of air quality (DEGLER, S. E.: *State Air Pollution Control Laws*, Bureau of National Affairs, Environmental Management Series, 1969).

The US Act concerned with the condition of air in cities and on highways does not mention expressly an olfactory nuisance, but is rather concerned with the toxic and pathogenic, especially with the carcinogenic, effects of the waste gases from cars and lorries. No motor vehicle is allowed to be on the road without having installed an afterburning device which makes sure that no vapours or aerosols be emitted from the exhaust pipes without having undergone complete combustion into water vapour and carbon dioxide, thus satisfying both hygienic and aesthetic requirements.

An Environmental Council advises the President of the United States on pollution control. It is hoped that such an advisory body will have the weight and power to reverse the deterioration of the environment; if not, a more active intrusion into the decision-making process at the appropriate level is foreseen.

Legislation regarding the pollution of water by discharging untreated waste into natural or artificial waterways, i.e., rivers or sewers, is only of recent date in USA. Until 1948, only the Rivers and Harbours Act, approved 3rd March, 1899, and the Oil Pollution Act, approved 7th June, 1924, both Federal Acts, were in force. The first generalized Federal Bill was enacted in 1948 under the official title of Federal Water Pollution Control Act of 1948, known as Public Law No. 845. This Act defines the pollution of interstate waterways as a public muisance, and also specifies, in a fundamental manner, the remedy by empowering the Surgeon General to suggest appropriate measures against the pollution to both the person causing, or responsible for, the pollution and the state in whose territory the pollution is perpetrated. Similar to procedure in England, the offender is given a chance to comply with official suggestion regarding the abatement of the nuisance. In the case of noncompliance, the Secretary of Health, Education, and Welfare (HEW) can request the Attorney General to bring suit on behalf of the United States to terminate the offence.

Under the terms of Public Law No. 845 all states sharing the waters of a single stream may form cooperative agencies for action in the area under consideration. They set up standards of water quality and act as advisers and instructors in questions of water pollution by waste effluents (162).

Another collection of anti-pollution legislation has been issued by CITEPA, the Centre Interprofessionnel Technique d'Etudes de la Pollution Atmosphérique (112). Information is provided on essential characteristics and clauses of legislation obtaining against air pollution in the selected countries named in the report.

Legislation is state and federal in West Germany. General enabling Acts have been passed in Belgium, France, Italy, but Holland has no specific anti-pollution legislation, and in Switzerland the Civil Code merely forbids nuisance. In France, it is the Loi du 2 August, 1961, sur les pollutions atmosphériques et les odeurs, in Italy Law no. 615 (introduced on 13th July, 1966, and in Lower Saxony, where there exists no federal law, there is the state law since 6th January, 1966.

Motor vehicle exhaust is especially legislated for in the United States by the Air Quality Act of 1967. The law affects those industries which emit dust (117). State and federal laws demand that the exhaust gases from cars shall not contain more than 1·5–2·3 per cent by volume of carbon monoxide, depending on the engine rating. Amounts of CO from passing cars at heavy traffic circles (11 000 cars/h) reach sometimes maximum allowable concentrations. From Japan, these figures relate to Osaka City (116):

$$CO_2 \quad 4 \cdot 6 \text{ ppm} \quad (\text{MAC 5000 ppm})$$
$$CO \quad 85 \quad \text{ppm} \quad (\text{MAC} \quad 50 \text{ ppm})$$

The effects of vehicle exhausts on human health have been often discussed. They have been blamed for lung cancer equally with inveterate cigarette smoking. But again, they have been described as merely a nuisance rather than a health hazard. Considering that petrol driven vehicles in Britain alone discharge five or six million tons of carbon monoxide annually, which the town dweller has to inhale together with unburnt fuel and a correspondingly reduced amount of oxygen, there is some substance in advocating a means of control. Even fifty years ago, there were similar complaints published, however, in a lighter vein.

> What is this that roareth thus?
> Can it be a Motor Bus?
> Yes, the smell and hideous hum
> Indicat Motorem Bum.

> How shall wretches live like us
> Cincti Bis Motoribus?
> Domine, defende nos
> Contra hos Motores Bos!

ALFRED DENIS GODLEY (1856–1925)
in *The Motor Bus*, 10th January, 1914.

Smell can be a legal issue in the activities of marine insurers and assessors who are called in by a shipper whose cargo has been tainted in transit by a residual smell from a previously shipped load (oranges, onions). The nose and the mind of the assessor, i.e., his experience, must guide him in the absence of stipulations of the law and indications of an instrument measuring residual smell, or 'taint'. As the matter lies at present, he will confirm or deny the claim on the strength of his degree of olfaction, and adjudicate compensation correspondingly. It seems that the real solution of this complex problem is a long way yet. An intermediate proposal might offer a working arrangement by elaborating detailed and specific definitions of 'taint' and limits of contamination which could be ascertained in every case by means of a specified chromatographic method.

Legislation on air pollution by smell is clearly a matter of national concern, at first. The case of marine insurance assessors involve proposals and agreements on an international basis. The International Standards Organization in Geneva might be able to help with this matter.

There are other approaches to the problem of bridging the serious gap in technical know-how of odour measurement without which legislation cannot formulate any laws or regulations concerning odour. It is up to the scientist and engineer to provide the means, and to act as advisers to the legislator. A common platform could be found by establishing osmogenic comfort thresholds which would reveal smell-levels of maximum tolerable concentration. These data would allow of formulating chemical specifications which would form the criteria in chromatographic determinations. On such a basis expert evidence will be acceptable to both parties, and limits of osmogenic pollution can then be defined by law, and verified by any qualified chemist.

Osmogenic comfort thresholds are not identical with odour thresholds. The former represent the upper limits of odour tolerance or endurance, and should be identified by testing people of various walks of life, different ages, and both sexes, but should be graded in accordance with certain occupations. Thus a comfort threshold for an animal waste processing plant, or for a pharmaceutical works, would be different from that applying to an office or a cheese-making plant.

A graduated scale of odour intensities would be required similar to the Ringelmann scale of smoke densities. This procedure has its attractions for the legislator who would have to specify a suitable odour comfort level of x to one, and of y to another, type of locality in accordance with the recommendations of his scientific advisers. This is less ponderous than using graded steps in mg/l or ppm in a legal text and would lead, in time, to handier instruments with scales giving relative readings.

Legislation for air standards also has had an interesting side effect on various industries which before were main contributors to air pollution. According to the National Bureau of Standards the estimated total air pollution in the United States weighs some 10 000 000 t per annum (164), and it is interesting to consider how legal action about it can have economic effects.

A typical instance is the case of the Central Electricity Generating Board, CEGB, of Great Britain. The quantity of sulphur, in the form of SO_2, put into the air from the Board's Generating stations was sufficient to have greatly ameliorated the crisis in sulphur supplies. The Board decided to recover the sulphur from the flue gases of their power plants and have built a prototype 10 t/day sulphur recovery plant into a 50 MW section of a coal-fired generating station with a total output of 1000 MW. The experimental work is being carried out in a Head Wrightson-CEGB plant incorporating a three-stage fluidized-bed reactor. Hot flue gas is entering the recovery plant at about 120°C, and sulphur will be extracted at the rate of 95 % efficiency. Alkalized alumina is the adsorbent used; it will be regenerated and returned for re-use. The process produces hydrogen sulphide which is further processed in a Claus plant by partial oxidation:

$$2H_2S + SO_2 \rightarrow 2H_2O + 3S$$

Petrol refineries use the same process to recover sulphur from oil. The refinery of the BP Company at Ingolstadt, Bavaria, recovers some 10 000 t of sulphur each year from the waste gases of the plant.

Besides the amenity advantages of such a recovery plant, it is very attractive economically. The capital investment is about 50 % below that of a sulphuric-acid-from-anhydrite plant; while at the time of writing the price of sulphur is high enough to justify almost any method of recovering it.

Thermal power plants using coal can be sources of radioactive pollution (235). A power plant burning solid fossil fuel will discharge up to 1 curie per annum of ^{226}Ra and ^{228}Ra per 1000 MW output. This contamination is in the form of radioactive fly ash. Oil burning plants will produce only about 0·5 millicurie per annum for the same output (239).

Coal is a prominent and most dependable source of sulphur (pyrites). It is only natural that some Local Authorities should be concerned with the amount of sulphur in the form of sulphur dioxide, in the atmosphere within their area, and that they establish sulphur standards for coal burnt in domestic hearths or industrial furnaces. This led coal merchants to attack the city fathers, in Chicago, for interfering with their trade in Illinois coal which is high in sulphur. Their Association asked the opinion of PROFESSOR W. MACHLE of the University of Miami School of Medicine who said that there is no direct evidence that SO_2 with or without fly ash is the sole or direct cause of illness, or that sulphur dioxide, in concentrations found in major cities the world over, is a health hazard (316). There is even a case on record where a supplier of coal, the Peabody Coal Co., filed a suit against the St Louis, Missouri, city authority, challenging the validity of the city's sulphur standard for coal (163).

12 Panels and evaluation

A panel of people trained in using any one of their senses with outstanding efficiency for the evaluation of physical or chemical phenomena may be considered an organoleptic measuring instrument. Such panels are of importance in the routine work of testing laboratories and as 'nose witnesses' in odour court cases.

Anyone who has worked on a panel, or has studied the methods employed and read panel reports, will admit that panel findings are subjective, and that even a trained panel member is not a constant adjudicator. While it cannot be denied that findings of a given panel may be consistent with one another, it is but rarely that findings of several different panels working on the same subject coincide to such an extent that deviations of findings and differences of opinion are outside statistical significance. Variations in panel response may be of such character or magnitude that they will mask the effects of their variables (85).

Whereas a measuring instrument will produce a reading objectively, any organoleptic assessment is bound to depend on physical, physiological, mental and—last but not least—pathological influences to which an observer is subject. To this must be added that confusion exists about the terminology used by panel members describing their sensory experiences. This inadequacy of terms applies to quantitative as well as qualitative descriptions.

The difficulties besetting anyone who is given the task of setting up a panel are formidable and can be outlined as follows:

(1) Deciding on the composition of the panel.
(2) Designing a training programme.
(3) Selecting individuals by preliminary tests.
(4) Designing the panel meeting and working room.
(5) Preparing standards of odour types in specified concentrations (84).
(6) Preparing a terminology.

(7) Working out procedure for all members at the beginning of a meeting.

(8) Designing a routine for presenting the samples to the panel.

(9) Presentation of results.

There are important observations to each item.

(1) *Composition of panel.* Panels of only four or five members are bound to produce erroneous or irrelevant data. When observations were made to establish colour preferences in man, more than 100 000 people were tested, and the scale of preferences in use today is universally applicable.

A panel for working with odorous substances should consist of physicists, chemists, physiologists, rhinologists, and a statistician. Or at least one of each group of professionals should belong to an advisory panel for guiding the members who actually test the material.

Other people joining the panel should represent a cross-section of ordinary citizens and be from all age groups, both sexes, and have as many different occupations as possible. They should be of all walks of life and embrace as many degrees of intelligence as possible.

Interpretation of nervous messages, such as those from the olfactory system, is a matter of mental activity; but even at the lower level of physiological functions diversity is immense. SHERRINGTON (45) explains the working of the nervous system thus: 'All parts of the nervous system are connected together and no part of it is probably ever capable of reaction without affecting, and being affected by, various parts.' Sensory perceptibility varies with age, sex, social background, environment, occupation, pathological conditions, to name just a few factors.

As to *age*, olfaction is at its peak at puberty or soon after, and begins to decline at about 45 years. Anosmia, the total lack of olfactory perception, occurs in 5 % of young people at 26, but gains with age, affecting about one-third of all people aged 78. If tests concern both smell and taste, as in the testing of foods, it should be taken into account that the number of taste buds in an old person has been greatly reduced, as are the flow of blood to the brain and the maximum oxygen uptake. Nerve conduction velocity is also reduced, and the last three data combine to result in memory loss, mostly of recent events—a fact of importance to elderly panel members.

As to *sex*, women's oxygen uptake is less than that of men. A woman between the ages of 20–29 years will require only as much oxygen as a man between 50–65. Oxygen is taken up by the blood in quantities reducing with age. Although the general intelligence is equal in comparable men and women, the latter have a higher verbal ability, power of observation, and immediate memory, whilst men are better at visualization.

As to *social background*, different strata of society will vary somewhat in the use of bathrooms and ventilation, although these demarcations are waning. Social background is however significant in that people get fatigued to those smells which constitute their permanent background; these become part of the background pattern which is no longer perceived consciously. If a person is now asked, as a panel member, to judge an odour belonging to that group he will not be able to produce a useful report.

As to *environment*, similar principles may apply. Tanners are unaware of the peculiar smell in their shop. Maggot breeders do not smell the stench of putrefaction or the specific odour of trimethylamine. Butchers and fishmongers are much less aware of any off-flavours their wares might have than is a prospective purchaser entering their premises.

The use of an electric fire in a room will cause the air to become dry and will reduce the cross-section of the nostrils.

As to *occupation*, a person highly trained in odour discrimination and recognition, such as a flavour chemist, or perfumer, is on the contrary a valuable addition to any odour panel, even if the group of odours to be tested is well outside his own usual occupation.

As to *pathological conditions*, there is, first of all, occupational olfactory fatigue. Hysterical anosmia will be but a rare occurrence in persons considered for panel service; it is caused by the strong irritation of the fifth cranial nerve (*n. trigeminus*).

Persons suffering from a cold in the nose may try to alleviate the complaint by instilling a nasal decongestant. A pathological, or congenital, swelling of the middle turbinate will cause a partial or complete blocking of the upper turbinate, resulting in a much reduced, or complete absence of, ability to smell. Ozaena, nasal allergy, and other conditions can have the same result.

Parosmia of a temporary type is often found in pregnant women who are, usually, quite unaware of the fact. On the other hand, olfactory acuity rises steeply prior to menstruation. This is a vestigial phenomenon and of interest from a phylogenetic point of view. The increase in acuity is of importance in female animals on heat which are eager to find their mate.

Temporary anosmia is a well-known characteristic of certain diseases. Recovery of the sense of smell may take a few days to several months. Recovery periods are 3 days for creosote, 11 days for skatole, 12 days for mercaptan, 90 days for caoutchouc.

Then there are substances to which some people are sensitive while others are not affected by them at all. This group of compounds comprises hydrogen cyanide, carbon monoxide, and others. This does not mean that threshold perception is high in one group and low in the other. The absence of odour acuity is a qualitative, not a quantitative, character.

In addition to physiopathological conditions there are psychopathological situations which affect the sense of smell. Imagined olfactory offences cause as much, if not more, trouble, inconvenience, and discomfort as a foul smell which really exists.

As to *more normal physiological conditions*, the effect of digestion on mental fitness should be considered. 'After dinner tiredness' lasts for about an hour after a moderate meal; if alcohol has been taken, much longer.

Generally, therefore, one should provide for an advisory committee of suitably selected scientists and experts to be at call; see that panel members represent a true cross-section of the population, and that the number of members and tests carried out by them conform to the requirements of the statistician. The ages of panel members should be in the range 20–50 years and a log should be kept for each member. Physiopathological conditions, upsets and mental stresses should be communicated by members to the team leader, who will be advised by his scientific panel about any action required. A generous lunch break of at least two hours should be allowed, but only for a moderate meal without any alcohol. Smoking should not be permitted on duty or during lunch hours. Members should be encouraged to go for a walk in fresh air during lunch time.

(2) *Training*. The training programme is concerned with explaining to the members the above recommendations; the importance of the tests; and arranging for a series of trial sniffings until coherent results are achieved, and the team leader can handle the panel like a sensitive instrument.

(3) *Suitability tests*. These are concerned with the handling of osmogenic compounds; they should also establish each member's ability to use the terms correctly and to estimate quantitative (degree of concentration) and qualitative (type of odour) characteristics. The members' olfactory abilities are tested with prepared series of odours (cf. 5), insisting right from the beginning that the vocabulary selected by the team leader be followed (cf. 6).

(4) *The venue for the tests*. Particular attention should be paid to wall surface structure, illumination, ventilation and quiet.

The *wall surface* should be very finely plastered, painted several times with a washable plastic emulsion paint. Ideally, all walls and the ceiling would be glazed, as glass has practically no odour-retentive powers. A plain oak or teak *floor* is of sufficient density to have little adsorptivity. The natural wood is covered with PVC sheeting or tiles; no floor polish must be used. Neither carpets nor curtains nor any other type of upholstery or fabric are used in the test room. It is best *painted* in a matt pastel shade of blue-green, with window frames and doors in, e.g., light brown or broken yellow—all colour tones should be such as not to distract attention. *Illumination* is through the windows, but it is psychologically best if the windows are blind, with artificial light sources hidden behind them. This will make for constancy in illumination level and colour, independent of time of day and weather conditions. This is a precaution against any person of the panel being affected in his olfactory acuity by visual perceptions.

Ventilation design must exclude draughts. For comfort in general the accepted figures are 30 m^3 (slightly more than 1000 ft^3) space per person, and 60–90 m^3 of fresh air supplied per hour and person. For odour test rooms these values are increased by 20–25%, as a minimum.

Central heating radiators must carry humidifiers; electric heating is avoided. The optimum temperature is 15°–20°C (60°–70°F) and relative humidity 60–70%.

An odour test room designed for panel meetings is, essentially, a *'clean room'*. The method of ventilation used in a clean room is critical and has been discussed in the literature (191, 193). The recently introduced laminar air flow method maintains a uniform air velocity and air flow pattern; the source is either an entire wall or the ceiling. Solid particulate matter down to 0·3 μm is practically eliminated (192). Standard temperatures and humidities in clean rooms should conform to US Air Force Technical Order 00-25-203, or equivalent specification.

Making the test room *soundproof* is recommended. Noise, like other extraneous stimuli, will distract attention from the task.

(5) *Preparing standards of odour.* This is a difficult task, even in relation purely to the working of the odour panel, and not to national and international specifications, which have yet to be defined in any case. For training purposes it is best to prepare two series of standard solutions, one being comprised of similar, the other of contrasting, odours. Each series consists of standardized degrees of concentration arranged in logarithmic steps in accordance with the Weber-Fechner Law. This stipulates that the increase in the concentration of odour (stimulus St) which is necessary to produce the smallest possible, i.e. just perceptible, increase in olfactory sensation S always bears a constant ratio to the whole stimulus. Fechner expressed this (Weber's) Law mathematically by writing $S = A \times \ln(St/St_0)$, where St_0 is the threshold of stimulus and A is a constant. Changes in concentration between 1:5 and 1:3 are obvious to most observers. Fechner's Law states that the sensation varies as the natural logarithm of the stimulus. The relationship between stimulus and sensa-

tion is shown in Figure 4. Weber's Law does not apply with very weak or very strong concentrations.

Changes in psychophysiological functions in man exposed to low concentrations of solvent vapours—trichloroethylene in this instance— were investigated in specially designed tests. At threshold and low concentrations there were no significant changes in psychomotor performance which declined progressively with increasing concentrations. Similar experiments were carried out with Freon 113 (1,1,2-trichloro-1,2,2,-trifluoroethane, $C_2Cl_3F_3$) confirming these findings (86).

(6) *Terminology*. The vocabulary of odours abounds in such descriptions as putrid, aromatic or fragrant, not even suited for generic formulations. General terms, by no means descriptive or characteristic, but regularly used in literature, in discussions, before Courts of Law, etc., include foul, pleasant, nauseating, acrid, fruity, repulsive, stifling, heavy, ambrosial, rubbery, heavy sweet, solventy (whatever that means), green sweet, oxidized, chemical sweet, dirty clothes, barn-like, moth balls, shoe polish, gauze-like, plastic, and many others. Comments are hardly necessary.

From the earliest beginnings of osmology the investigators tried their hand at defining a classification of 'basic' odours; here are a few:

CROCKER and HENDERSON (46): 4 basic odours—fragrant, acid, burnt, caprylic.

HENNING (47): 6 basic odours—spicy, flowery, fruity, resinous (balsamic), empyreumatic (burnt), offensive (foul).

HEYNINX: 6 basic odours—spicy, vanillic (ethereal), acrid, alliaceous (garlicky), empyreumatic, foetid (rotten).

ERB (48): 6 basic odours—acid, fragrant, empyreumatic, caprylic, infra-odours (strong and pungent), ultra-odours (too weak to register).

ZWAARDEMAKER (49): 9 basic odours—aromatic, balsamic (fragrant), ambrosial, ethereal, alliaceous, empyreumatic, caprylic, repulsive, foetid.

RIMMEL, quoted by MONCRIEFF (19) used even 18 basic odours—Rose, Jasmine, Orange Flower, Tuberose, Violet, Balsamic, Spice, Clove, Camphor, Sandalwood, Citrine, Lavender, Mint, Aniseed, Almond, Musk, Ambergris, Fruit.

This list is quoted to show that not every classification is formed on a general concept. This is plainly a list of scents intended for the perfumer. It includes neither offensive nor food odours.

Since the pioneer work of ZWAARDEMAKER (49), almost every worker in the field of osmogeny has his own views on the subject—and his own system.

Odours require both qualitative and quantitative description, based on fundamental concepts, and applicable to any odour existing or created by future research. Where variety in perceived stimuli is great, the tendency is to seek 'fundamental' concepts. This is a sensible step, but one fraught with difficulties which seem insoluble.

Ideally, 'fundamental' odours would be such that mixtures of any of them would produce any odour known to man. CROCKER and HENDERSON (287) propose to attribute nine degrees of intensity to each fundamental odour, and to describe any odour by a four-digit number (each number representing one of the nine degrees of intensity) and the order of the digits describing, by their sequence, the fundamental odour to which the intensity digit refers. The standard arrangement is, moving from left to right,

first digit   fragrant
second digit   acid (sour)
third digit   burnt
fourth digit   caprylic

Thus, 6423 is the smell of the rose consisting of

a fragrant component (first digit) of intensity   6
an acid component (second digit) of intensity   4
a burnt component (third digit) of intensity   2
a caprylic component (fourth digit) of intensity   3

Classification 0000 means complete odourlessness. Between 0000 and 8888 (0–8 are nine degrees of intensity) any combination, i.e., a total of 9999 can be expressed by the scheme.

A more precise method of describing odour intensity (225) is the ASTM method of test D 1292. This elicits the number of times the osmogen is diluted by a factor of 2, i.e., to one-half its original concentration c, to reach osmogenic threshold θ. Then $\theta = c/2^n$, where n is a number called the Odour Intensity Index (OII).

There are other odour indices, for instance:

Odour index	Representing a concentration odorous substance lb per 10^6 ft^3	Description
0	0·001	imperceptible
1	0·01	threshold
2	0·1	definite
3	1	strong
4	10	overpowering
5	100	dangerous

where log conc. might give another index:

Odour index	Log conc. odour index
0	-3
1	-2
2	-1
3	0
4	1
5	2

A research project carried out at the Illinois Institute of Technology (288) is concerned with the numbering of smells, i.e., registering odours by their qualitative and quantitative characteristics, and also hopes to discover the secret of designing instruments for measuring the quality and quantity of odours.

(7) *Procedure*. During service on the panel, members should be required not to use any personal perfumes for any purpose. Test substances should not be prepared in the test room, and odorous substances must not be spilled, or be allowed to evaporate unchecked.

Upon arrival, the members should be conditioned for 20–30 min by breathing 'pure' air. This time will suffice to clear their mucosae from any particulate matter. After that, a blind test is useful. It will indicate any inherent sensations, real or imagined, and the results are noted in the personal log books.

Comparative tests might be advised with a smaller panel using one, then both nostrils for inhaling. When both nares are used, lower threshold values may be expected. Members should be examined before continuous and long series of tests for sinus trouble, stopped-up nasal passages, adenoids, and other obstructions to, or irregularities of, breathing.

A further point to watch is the time required for the olfactory epithelium to match itself to the stimulus. Too few data have been published on this aspect to have statistical significance. In the absence of definite numerical guidance, an application time equivalent to one full inspiration should be permitted, i.e. about three seconds per deep breath.

A layer of surface moisture must bathe the sensory membrane in the nose. It is imperative that the full volume of inspired air should pass across the olfactory epithelium behind the root of the nose. The technique of breathing-in the test air must, therefore, be adapted to attain this aim. A sharp sniff ending in a deep breath is ideal. At very low concentrations, especially at sub-threshold experiments, repeated sniffing will be necessary to bring the required number of molecules to the sense organs. In between sniffs, the air should be exhaled through the mouth, not through the nose, in order not to disturb the cumulative effect of repeated breathing.

Upon arrival, the members should be conditioned for 20-30 min by breathing 'pure' air. This time will suffice to clear their mucosae from any particulate matter. After that, a blind test is useful. It will indicate any inherent sensations, real or imagined, and the results are noted in the personal log books.

As an olfactory sensation can be produced only if sufficient molecules of a substance are swept up into the upper reaches of the nasal cavity by means of a short sharp sniff, they can be removed from the site of action only by a similar sniff working in the opposite direction. Just as it may not be possible to gather sufficient molecules by a single inhalation, a great concentration of molecules in the area of the third turbinate may require several strong expirations in order to blow them all out, and clear the nasal space of them.

Refractoriness and fatigue are two conditions affecting the acuity of the olfactory system. It is therefore recommended to keep the rate of application of osmogenic stimuli fairly low so that the nervous system can recover. Equally, it is best to apply contrasting odours at different sessions, or one kind in the morning, the other during the afternoon meeting. Spacing them closer together in time is not recommended.

(8) *Presenting the samples*. The substance giving off the odour is contained in a small bottle which ends in a stopcock. Between this and another stopcock lies a volume of 1 cm^3. The concentration of the odour is known to the manager of the panelists, and should be very low. A series of such bottles, each containing a different concentration of the same substance, is prepared. The far stopcock ends in a short piece of glass tubing to which is attached a piece of rubber tubing suitable to be introduced without difficulty into one nostril of the panel member. If no such adapter can be used because it has an odour of its own (rubber odour) a piece of silicon or nylon tubing should be used. Very thin PTFE tubing would be ideal but it may not be found sufficiently pliable for the purpose. When introducing it into the nostril, no lubricant must be used, the adapter must not cause any irritation to the mucous membrane in the nose, and the fit should be fairly tight, but always comfortable. The panel member should not become conscious during the test of the adapter in his nostril.

The threshold of a particular odour relating to a particular person is reached when the test subject, after inhaling sharply through the nostril fitted with the adapter, cannot register an olfactory sensation at all. It is of questionable benefit of the tests to discriminate between the sensation of an odour which is so weak (sub-threshold) that it does not allow of identifying its cause qualitatively, but only registering its presence, and an odour which is just discernible by its characteristic note (threshold value). The mere sensation of the presence of an olfactory stimulus without the possibility of realising its quality is of no consequence. As no quality is identifiable, there can be no discriminating the sub-threshold of smell A from that of smell B. The olfactory sub-threshold is similar to the visual sub-threshold at which the eye can perceive 'light', but neither shape, nor dimensions, nor colour. Here, too, identification is impossible at sub-threshold values.

The panel should be told the application of the test substance, for instance, whether the odour will be incorporated in a perfume or cream, or a piece of clothing, or a food or drink; or whether the odour in the concentrations presented to the panel is considered offensive, nauseating, agreeable, evocative of certain feelings, etc.

They should be instructed to use no terminology other than that prepared for their use, or accepted as a national or international standard. Each test should be repeated three times without the member knowing which substance he is going to judge next. Large deviations from the consensus of opinion and evaluation should be investigated to find the reason for their existence. It is essential to eliminate this reason. Small variations, say 5-10%, may be neglected.

(9) *Presentation of results.* It is the panel chairman's task to summarize the findings in a brief report for the management. Acceptance or rejection of an odour, or its characteristics (agreeable, obnoxious, etc.), or its suitability for a proposed application should be expressed as a percentage value of all opinions. The report must not contain opinions of members or the chairman other than those appertaining to the set task.

Further research on evaluation

There is a search for a 'definitional model' of public perceptions of air pollution (205). A survey has been conducted in Southwest Pennsylvania, and the records were analyzed in terms of education, residence, socio-economic status, sex, and other social characteristics.

A variety of designs have been proposed for a test facility. An installation for the testing of large numbers of people requires a mobile laboratory and a large bus (10 m long) was selected (206) affording six identical face exposure chambers. The system of odour generation provides suitable concentrations ranging from 1 ppm down to 1 ppb, or less. Each test person moves through all six chambers consecutively where different concentrations of an odour are sampled. The sequence and degree of concentration is random and unknown to the subject. An exact terminology consisting of only three words, is prescribed. The person reports his sensations as 'detected', 'pleasant', or 'unpleasant'. Personal data are then provided by the individual such as age, sex, occupation, smoking habits, abnormal olfactory perception, and others. The data are then transferred to punched cards and analyzed to show if there is a significant influence upon the response to odour intensities and individual reactions.

It is realized that quantitative measurement becomes the more important, the more elusive the qualitative instrumentation seems to be. Most systems are based on the idea of quantitative dilution, some following the Weber-Fechner law, some being a decimal scale dilution in equal arithmetic steps, some others following random concentrations. The touchstone is the human vote, perhaps in an improved form as outlined above. Organizers of tests seem to prefer large volume test quantities to neutralize the losses of odorous material on the walls of vessels, but it seems strange that some of them are content to use as few as two or three subjects (207).

13 Odour evaluation in court cases

Depositions and statements made by witnesses in or out of court vary to a considerable degree from one another. It is a well known fact, often demonstrated during lectures on psychology, that no two people will give the same detailed account of an event which is being performed in front of them for their benefit. This does not only apply to eye-witnesses, but even more so to nose-witnesses, because man is much more trained in using the visual sense consciously than the sense of smell. Differences in statements have their reasons not only in mental capacity, such as power of observation, but also in physiological, and often morphological variations of fundamental elements. This peculiarity is instanced by the many and varied odour threshold values published by various workers. The following data are taken from investigations by Robert A. Baker of the Franklin Institute, Philadelphia (165).

Substance	Number of panellists	Number of observations	Threshold odour level (ppm) Average	Range
Acetophenone	17	154	0·17	0·0039– 2·02
Acrylonitrile	16	104	18·6	0·0031–50·4
n-Amyl acetate	18	139	0·08	0·0017– 0·86
Pyridine	13	130	0·82	0·007 – 7·7

The column commanding attention is the last one. The range of threshold values in individuals varies, in these four cases, and especially for acrylonitrile, between 1:500 and 1:16 000. The average odour threshold becomes a pretty meaningless figure when it is considered that the average is 18·6 ppm for acrylonitrile, yet at least one panellist was capable of smelling it in 0·0031 ppm concentration, whereas at least one other member of the panel was incapable of ascertaining the presence of acrylonitrile in air until the high concentration of 50·4 ppm was reached. This is not only about three times average threshold concentration, but also more than twice the maximum acceptable concentration (MAC = 20 ppm).

From the legal point of view this position can involve judge and jury in inextricable and interminable difficulties and discussions. Assuming that a test case is brought and the question of 'smellability' arises; depending on whether witness A (threshold 0·0031 ppm), or expert witness B (50·4 ppm) will testify for the prosecution, the defendant—allegedly causing a nuisance at 18·5 ppm—will lose or win his case.

However, the judge may have been interested in problems of olfaction, and have read something on the subject. He will be aware of the wide range of thresholds of osmic perception, and he will also know that there are 'average' thresholds. He will therefore delve still deeper into the subject and get from his library standard works on olfaction. Looking up the chapter on thresholds, he finds that nearly each reference work lists tables of widely divergent values. He brings up this point at the hearing, asking the expert to enlarge, and advise, on this point. The expert speaks very knowledgeably of differences in methods and techniques of organoleptic tests, of the many hazards which may beset a panel member without his being aware of any change in olfactory faculties. Environment also comes into play and—the judge has realized it by now—there is not really a sure, dependable way of guiding the court objectively unless the judge would prefer to accept any one average value as a basis for his dictum. This, the expert says with a wry smile and the judge nods agreement. This invites objection from both sides, the prosecutor losing his case, because the judge happened to chose a high average threshold, or the defendant losing his case because too low an average, in his opinion, was quoted by the judge.

The wide discrepancies in listed values together with the impossibility, at present, of designing practical olfactometers have a bearing on legislation about acceptable odour limits. A Smell Abatement Law cannot be made operative without specifying limits of concentration. Two other approaches are possible: available data of major importance are considered by committees of experts, and selections are made as a recommendation to the legislature; or a working committee is charged with the task of specifying a standard method of olfactometry which should be considered internationally, and be made the basis for new threshold measurements. It is thought that the new measurements should be made on a national, not an international, basis as the newly established values will mirror racial and national olfactory characteristics. On scientific advise, the legislator will then stipulate a limit of osmogenic air pollution, going beyond which will make a person automatically liable to prosecution without having to establish in a lower Court whether a nuisance has been committed. Instruments will then be available for repeatable measurements which no longer will be based on organoleptic faculties.

14 Anomalous odour perception

Just as there are people with anomalous sight or anomalous hearing, so there are a significant proportion having anomalous olfaction. This is caused by either anatomical peculiarities as, for instance, no odour perception at all—complete anosmia—which would range with complete blindness or deafness, or it may be caused by the absence of some section of the odour 'spectrum' known as partial anosmia, both of which may be congenital, but the latter may be also acquired as a kind of involuntary or voluntary reaction to the environment (merosmia). When only objectionable odours are perceived, and nothing else, one speaks of parosmia. Considering that odour as such, i.e., outside the person, does not exist, but is a result of mental evaluation taking place in the higher centres of the cerebral cortex, parosmia is not to be regarded as a special case of merosmia. Parosmia is, rather, a mental reaction to an experience in the past.

Osmodysphoria, the dislike of certain smells not generally considered objectionable is the reaction to a personal unpleasant experience caused, or accompanied by, that smell. There is the generalized dysphoric principle which leads a person to dread all odours (osmophobia). There is also a very rare condition known as olfactism. This means that a sensation of smell can be produced in the person subject to olfactism by other than osmic stimuli. A sensation of odour may be induced in a person so adapted, by a visual signal, or by sound. Olfactism is, functionally, similar to chromaesthesia, photism, and collateral excitation of other kinds, referring to the senses. Olfactism can be taken as proof that osmic perception is nothing but a product of cerebral activity functioning under specific, though abnormal, conditions. A possible explanation is as follows: There are two different sensory receptors working, for instance, one olfactory, and one visual sensor. There are also two separate medullated nerve fibres each connecting its own receptor with the brain. (A medullated nerve fibre behaves like an insulated metal wire, a greyish-white substance forming the conducting inner fibre, the white myelin forming the insulation.)

An energy pulse travelling along the medullated nerve fibre cannot, normally, cross over to another nerve fibre because of the insulating property of the myelin sheath. In the case of an existing collateral excitation it is possible that a single fibre has a congenital or acquired 'insulation failure' thus allowing the pulse conveying a single message, e.g., 'blue light', to leak its energy (the message) to a similarly damaged fibre of the olfactory system. This will produce a specific coexcitation whereby the seeing of a blue colour or light will excite both a sensation of blue in the visual area of the brain, and a sensation of an odour in the olfactory cerebrum. It may be an odour of roses, or sewage, or anything else. For the olfactory nervous system, the leaked visual energy pulse may, but need not, be specific. General coexcitation exists if any (visual) stimulus involving insulation failures in several optic fibres will cause the sensation of an odour. Multiple coexcitation refers to the fact that insulation damage is rather on a large scale, stimulation by any colour producing the adequate visual signal together with the perception of several, mostly non-specifiable, odours. Extrasensorial perception of odours has not been described in the literature and may not exist at all.

Part III

INDUSTRIAL SOURCES OF ODOUR POLLUTION

15 Meat and fish processing

The services rendered by the 'offensive trades' to the community are essential, and fundamental to maintaining hygienic conditions. Without them, garbage would ferment in the streets, sewage would pollute streets and rivers, butchers would have maggot-ridden offal and bones in their closets, and all other wastes of biological and metabolical origin, of trade and industry, would make habitation an ordeal on a vast scale. Perhaps these thoughts will contribute a little to seeing offensive trades in a different, more realistic, light.

The list of offensive trades published in UK Statutory Instrument 1950, No. 1131, is truly impressive. Those mentioned in Classes IV and VIII are of particular interest here as they include such trades as blood boilers, fat melters, maggot breeders, chitterling and tripe boilers, fish processors, glue makers, and some of the most important sections of the chemical industry, for instance paint, varnish, and resin manufacturers; but synthetic resins are an exception.

The trade premises vary in size and quality from an old tumble-down shack or Nissen hut on, perhaps, half-an-acre of waste ground at the outskirts of town, to a modern factory built within parkland in easy reach of the community. The small place causes more trouble to residents and the Local Authority than any number of large factories. There are, of course, exceptions to this statement, but where a large chemical plant causes a smell nuisance, it is either the main product, or an inevitable and valuable byproduct which may give trouble by accident. Even the technically most advanced production system is not free from occasional breakdowns: the failure of a valve may cause large volumes of highly offensive gasses in concentration, to be released to atmosphere.

The rendering of animal skins, bones, and soft tissues in the production of glue (an impure gelatin) causes offensive odours owing to the sulphur content of the products. Gelatin, an impure albumin of the general formulation $C_{76}H_{124}O_{29}N_{24}S$ to $C_{102}H_{151}O_{31}N_{39}S$ is graded into tendon, ligament, and commercial gelatin in order of rising sulphur content (0.26, 0.57, 0.71%) and increasing offensiveness. The making of glue is a hot process and the atmosphere in such works is steam laden. Air treatment is always difficult and expensive and, therefore, rarely carried out.

A significant improvement was made in the processing of bones by the invention of a process using shock waves transmitted through cold water for rupturing the cells and releasing their fat (Chayen process) (156). The degreased bone is then dried and ready to be further processed. It is the basic raw material for animal feeding stuffs, bone phosphate, calcined bone for the production of china, fertilizer, gelatin, and glue. The extracted fat is for the making of greases, tallows, and other products.

This cold process, patented to British Glues and Chemicals Ltd, is carried out in modern factories which are the first ones of their kind working completely free from any odour nuisance inside or outside the works. Polluted air is drawn through furnaces where the easily combustible osmogens are burnt. Oxidation of the other osmogenic pollutants takes place in specially designed hexagonal cylinders fitted with ultraviolet ray generators. The radiation, forming nascent oxygen, is contained within the hexagonal vessel, and after passing through the curtain of radiation the purified air is released to atmosphere, Figure 8. No ozone is used within the factory building, only within the confines of the ventilating system. The whole building is maintained under a slight negative pressure so that air cannot leave by any other way than that prescribed by the ventilation system.

The rendered, dried, and ground waste from abattoirs is used as fertilizer or feeding stuff and goes by the name of tankage. Its components include: N $5-10\%$; $Ca_3(PO_4)_2$ $8-30\%$. The dry rendering of animal carcases is a hot process and still carried on in most cases in the old-fashioned way which gives offence even to not so sensitive nostrils. The carcases arrive either from the abattoir where they have been condemned, or from other sources. Waste from butchers is also collected and brought to the works mostly in open trucks covered only with an ill-fitting tarpaulin. During the warm season, this load attracts blueflies, and the material is blown even before arrival at the processing station. Once there, it is unloaded on to a platform below which there is an array of steam jacketed boilers. It is not unusual for the raw material to remain on that platform for one to several hours depending on the availability of cookers. Cooking time varies between 2 and 6 hours. Even the best planned collection will lose time owing to unforeseeable circumstances, and a load of material may arrive after working hours and remain on the platform until the next

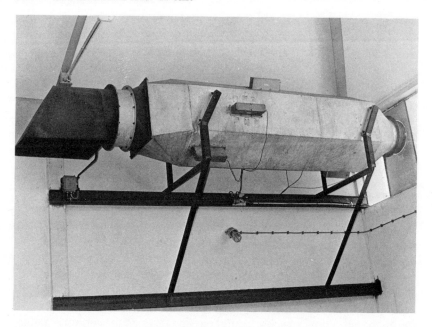

Figure 8   Extraction unit which purifies the air from a 'Chayen process' bone fat extraction plant, using ultra-violet radiation for destructive deodorization. Polluted air enters the extraction unit from the left and is released, deodorized, directly to atmosphere (right). The ultra-violet generators can be seen on the outside walls of the hexagonal unit. (*Courtesy British Glue and Chemicals Ltd.*)

morning or, if this happens at a weekend, until Monday morning. The condition of it by the time the works start-up again is beyond description —and so is the smell which has been rising from the carcases for 12 to 36 hours. The usual odours of putrescence are enriched by a good portion of trimethylamine produced by the biological process of maggots feeding.

It is not so difficult to prevent these conditions, even if the waste products are collected from a large area involving long journeys. The open trucks should be boxed in so that flies do not find their way so easily to the meat which should be carried in individual and closed palankins. Collecting rounds should be so timed as to make sure there are no long waiting periods at the works before the material can be loaded into the boilers. It is less costly for the processor to have one of his cookers stand waiting for, perhaps, half-an-hour than have a smelling load lying on the platform for 4–5 hours.

The discharging of the boiled material is another cause of serious smell offences. The hot mass of cooked meat, bones, hooves, is discharged into trolleys which carry the product to the centrifuges for the separation of the fat from the remainder. This is then used for cattle food, fertilizer, and other applications. The hot cooked waste gives off all types of aliphatic amines (open chain carbon compounds to which is attached the amino, NH_2, radical), sulphides, aldehydes and trimethylamine. Fresh meat cooked produces hydrogen sulphide only in small quantities, and practically no trimethylamine; but it is nearly always stale meat which goes through the cookers and this will produce more than twice the amount of hydrogen sulphide, besides trimethylamine.

Some persons can endure the strong 'meaty' smell, others find it revolting. To deal with premises of waste processors an effort must be first made to put roofs and walls in a good state of repair, and to enforce the closing of doors and windows. These are habitually kept open for 'ventilation'. Cleanliness is essential. Floors must be kept free from grease and fat and unprocessed raw materials should be prevented from getting covered in maggots. In the stores where greaves or fertilizer meal is kept in bags or in great piles there will be accumulation of greasy dust (fibres, fertilizer, dry blood) on structural steel and walls, roof trusses, crevices, window ledges, and unless all this dirt is cleaned away regularly, there is no chance of controlling air pollution. The next step is to design a ventilating system which will collect the air; first of all by erecting hoods above the direct sources of odour, and connecting them to the ducted system, then—if necessary—by arranging for another system to collect the general air in the usual way by another ducted system. The 'production air' will pass through grease filters which must be effective and adequate, and must be cleaned at least once every day; otherwise the ducting will soon become coated and after a time blocked, by the grease condensing and solidifying. Grease traps may be used in front of the filters. A powerful deodorizing action must then be introduced before the air can be released. Such a drastic method is not required for the 'general air' which will need but little deodorization. Two ventilating systems are suggested for locations where there is a great difference in odour concentration in the various sections, and extracting all the air in one common duct would make the system too big and expensive.

The floors should be hosed down continually. This may mean an extra load on water supplies as well as sewers and, here again, pretreatment may be required by power of the Public Health (Drainage of Trade Premises) Act, 1957. General conditions for accepting a trade effluent are a maximum temperature of 110°F (about 44°C), a pH value between

6 and 12, and suspended solids not in excess of 500 ppm. Each Local Authority can supplement those requirements as necessary. Special regulations are in force for toxic effluents.

Chlorination of trade effluents, whether osmogenic or not, will make them acceptable to the Authority. A closed water circuit can contribute considerably to savings and reduce the load on water supply and sewerage system. With washing towers either chlorine or chlorine dioxide is used.

The sources of smells generated in works processing animal carcases, and factors affecting the emissions, are discussed in detail, and from the works manager's point of view in an official publication by the Institute for the Control of Immission and Land Utilization in the Province of Nordrhein-Westphalia (198).

It is standard practice to process the various types of animal tissues in separate works: soft tissues, i.e., meat; osseous tissues (bones), and keratinous tissues (horns, hooves, hair), are always treated separately. Skins go to tanneries. Smells caused at the various works differ greatly in quality and quantity.

Meat processing works are the ones most likely to receive stale and flyblown material. The smell is one of hydrogen sulphide and amine. Bone processing works receive material which may be stale, but is rarely fly-blown, because the bones are well scraped as a rule, and bare of any meat. The cold processing of bones is odourless. Only the old type bone grinder taking his daily requirements from a large heap of old and evil smelling bones outside his works will cause a nuisance. Horns, hooves, and hair are a material with cosiderable amounts of sulphur in the cystine molecule

$$S.CH_2.CH(NH_2).COOH$$
$$\|$$
$$S.CH_2.CH(NH_2).COOH$$

which breaks down during boiling to yield sulphydryls methyl mercaptan CH_3SH, ethyl mercaptan C_2H_5SH, and hydrogen sulphide H_2S. The main substance of hooves, horns and hair is keratine $C_{41}H_{71}O_{14}N_{12}S$, with about 10 % cystine. The amount of sulphur in horns is about 3·20 %, whereas gelatin yields only between 0·26 and 0·71 % sulphur. Qualitatively (mercaptans) as well as quantitatively (3·20 %) the processing of kera-tinous matter is the most offensive process of the three.

Mercaptans are best made inoffensive by treating them with nascent oxygen when the sulphydryl radical is substituted by a hydroxy group, for instance, $CH_3SH + O \rightarrow CH_3OH + S$, or generally,

$$C_nH_{2n+1}.SH + O \rightarrow C_nH_{2n+1}.OH + S$$

16 Tanneries

Odour problems in tanneries are particularly disagreeable if the tannery is situated in the middle of town. This was not unusual a century ago when there was little through traffic and the indigenous people found nothing wrong with a smell which had always been with them. But changed conditions make it unacceptable.

Slightly different is a position typical of many a locality in England, and does not apply only to tanneries, but also to all those trades which were formerly carried on in rural surroundings, like sewage works, tallow melters, bone grinders, and others who, suddenly caught in the whirl of development, saw new houses spring up one by one, or in groups, extending from the rural town towards their works until the newcomers lived right on their doorsteps. The people who purchased these houses were always told of the existence of an offensive works nearby, but they were so eager to get a home that they pretended there were no smells. Once they had moved in, they suddenly became hypersensitive to odour, and began to complain to their Public Health Inspector. Life was now difficult for all three: the Local Authority, the owner of the premises, and the people living next door to them. 'The smell' was used in obtaining a reduction in rates; or to bring court cases against the trader; and unhappiness was complete when the Courts decided in many such cases on the principle 'who was there first'.

Yet, the smell from a tannery is not, or need not be, offensive. The prevailing type odour is of tannin and leather. None of the three groups of tannin, the pyrogallol tannins, catechol tannins, and phloroglucinol tannins, have an offensive smell of their own. To remove flesh and hair from the hides, they are painted with a mixture of lime and sulphuric acid. This process is not osmogenic either. A very offensive smell however does arise from maggot infested skins, especially before they reach the pupa stage. For obvious reasons, it is the heap of waste from the tanning process which becomes fly blown, develops heat, and a repulsive smell. Here, as with many offensive trades, an orderly conduct of the premises will prevent the nuisance from establishing itself. Waste is unavoidable, but it must not be deposited in the open where degradation processes occur uncontrolled. This is especially bad at times when the tip is disturbed for loading on to a vehicle for disposal. Maggot infestation is strong in spite of the presence of much 'paint' on the waste.

Spraying with a masking agent is of no avail because it cannot tackle the cause of the smell. If it is not possible to use large closed containers (palankins) and keep them closed, or have the waste carted away daily, then the only effective help is the irradiation of the waste material with ultra-violet rays which discourage flies from settling on the material, kill any maggots that have developed in marginal locations, and keep the waste covered with a layer of nascent oxygen. This will act as an asphyxiant for insects, flies, maggots, etc. The rays from the generators should cover the heap as well as an ample marginal area some 4–6 ft around it.

Tannery waste has one of the highest BOD (biochemical oxygen demand) values: 3500, which is equivalent to waste from 20 000 people. The standard value of BOD is 0·167 lb of oxygen required to treat the metabolic waste per head of population.

17 Tallow melting and refining

Tallow is the generic term describing solid animal fat from all sources. Its main constituents are palmitates, $C_{15}H_{31}COOR$, and stearates, $C_{17}H_{35}COOR$ of glycerol, $C_3H_8O_3$, with some oleates, $C_{17}H_{33}COOR$, of glycerol. The melting point of tallow is 45°C, varying with composition. At temperatures below 45°C, there is hardly any offensive smell except for that of cooking meat with a strong note of fat. The blue fumes rising from fat heated to above 45°C are indicative of the formation of acrolein and other burnt and carbonized products which are offensive and irritant. If a hood and ducting is used the blue vapours will enter them hot, but will soon condense on the colder walls of the duct and clog the ducting which may even be completely blocked at its mouth behind the hood. Condensing, or scrubbing the vapours close to the mouth is an essential requirement, though rarely applied. Grease filters should be used between hood and duct mouth. They will deter the onset of the clogging, but must be cleansed often. The condition in duct and ventshafts carrying hot fatty fumes is given away by fat tears oozing from joints. It is quite clear that any accumulated fat cannot be removed from the ducts, and forms a secondary source of air pollution in itself whether air is passing through the system, or not.

Tallow melting and remelting, a cruder way of purifying it, will always put a heavy load on the drainage system. Grease traps are mostly the only means of retaining the fat and preventing it from getting into the sewers where the congealed mass causes serious obstruction. Any efforts to retain and recover the fat are of economic importance.

Tallow is refined by washing in benzine (explosive) or tetrachloroethylene (perchloroethylene), C_2Cl_4 (non-explosive). The process is carried out in closed tanks connected together and to the pumps by suitable pipework. Normally, there is not much odour released, but when strong (not necessarily bad) smelling tallow is processed, fumes in concentration may

111

require treatment. These gaseous effluents often still carry fat globules and are likely to contaminate air deodorization equipment.

Similar processes are sometimes used in the extraction of fat from bones. Very little odorous matter is produced in the impulse method of cell rupture by shockwaves breaking open the cells of bones immersed in cold water, and blasting out the fat.

18 Gut cleaners

Gut of slaughtered, eviscerated animals is stored in small tanks of the dustbin type, and collected once daily from the abattoir. The method of storing gut in bulk causes the temperature of the entrails to rise considerably during several hours' waiting time which, in turn, sets the contents of the viscera fermenting.

When the guts eventually arrive at the gut cleaners (casing makers, sausage skin makers) they are spread on the floor and their contents are emptied by using water jets. The gaseous decomposition products—the guts are well and truly expanded to many times their diameter—are released producing a severe pollution of the air. From then on, the process is inoffensive, the guts being thoroughly cleaned, the fat is scraped off, they are salted, and packed into bins for storage or further processing. The smell from guts being emptied of their contents gets worse as one passes from beef guts to those of pigs and sheep, in that order.

Local bylaws usually provide for all windows at a gut cleaner's premises to remain closed, and doors to be similarly treated. Sometimes, air locks are a condition. It is not possible to separate process from environmental air. Masking is of no avail, combustion has been tried with but little success because of the low temperature produced in the furnace owing to the ingress of the large volume of cold air, and wet-scrubbing leaves residual odours.

Ultra-violet irradiation is completely successful and, by cautiously controlling the inevitable production of ozone to below its MAC-value, suitable and acceptable working conditions can be produced and maintained. A gut cleaner's premises represent one of those rare cases where ultra-violet generators must be used in the work rooms—not in ducts—and only a small number of them will be required as a rule. They should be switched on or off as conditions demand. Eyes, face, and hands, especially when they are wet (as they nearly always are in these shops) should be

113

well protected from the direct impact of ultra-violet rays. The water droplets act as focusing lenses and concentrate the otherwise harmless radiation on a small area on the skin to the detriment of that spot. Coloured skins are as sensitive as white skins.

19 Tripe boilers

Tripe is the principal part of the stomach of cattle, used as food. The stomach consists mainly of a muscular outer wall which is capable of churning the ingested food, and of an inner lining which secretes pepsin and hydrochloric acid and other substances, summarily called 'gastric juice'. It is the innermost lining (mucous coat or membrane) which, when separated from the stomach wall, cleaned, and boiled is sold as a food under the name of tripe.

The cleaning process removes any remains of food the animal has taken before slaughter, and destroys the pepsin, hydrochloric acid, etc. This is done by boiling the mucous membrane until it is a soft spongy texture and of a white colour.

The tripe boiler uses either open vessels, sometimes covered with an ill-fitting lid or sheet of metal or, working on a factory scale, steam cookers operating under normal atmospheric pressure, or using a light positive pressure. There is no offensive smell from a piece of tripe being cooked. The smell gets very powerful when a few thousand pieces of tripe are cooked at once and the boiler—usually of the vertical type—is opened at the end of a cooking period and the contents are discharged. Odour abatement is certainly required, but achievement is difficult.

Masking is inadmissible, because the spray will settle everywhere in the place, including the tripe, and this could be classed as food adulteration if a more serious view from the hygienic and medical point were to be preferred. Ultra-violet destructive deodorization would require ancillary services which would raise the cost of the installation. As a great deal of steam escapes from the cooker upon discharging its load, a wet-scrubber or condenser is required to remove the steam which is opaque to the rays. There is almost never a ducted ventilating system installed in these premises, so this would necessitate another rise in the cost of air deodorization.

However, there is one solution to the problem, i.e., to allow the boiling

115

vessels to cool down every time before they are opened for discharge. This method will cause a small increase in the overall running cost of the establishment because of the time loss involved in cooling down a hot cooker and getting it up to operating temperature again after each new charge. These costs are infinitesimal compared with the expense of installing a ducted ventilation system, ultra-violet generators, and an expansion chamber.

20 Processing of white fish

The processing of herrings for animal protein, fish oil, and the remainder for fertilizer employs a considerable number of people, mostly in seaside towns and villages where this industry has been thriving for long periods.

Fish tankage (fertilizer made from fish offal and caught, but not processed, non-edible fish) contains 6–10 % N, 0·4–8·0 % P_2O_5. *Animal tankage* (fertilizer made from slaughter-house offal) contains 5–10 % N, 8–30 % $Ca_3(PO_4)_2$. *Garbage tankage* (fertilizer made from household food waste) contains 2·5–3·5 % N, 2–5 % P_2O_5, 0·5–1·0 % K_2O.

Offensive odours always have existed in such areas and, if anything, were worse fifty years ago than today. However, since then people have moved into these towns and, being outsiders, and not used to fishy smells, have complained about it ever since. Generally, it must be conceded that people have become increasingly smell-conscious, and conscious of their civic right 'to the comfortable and healthful enjoyment of the premises they occupy, whether for business or pleasure' (A.-G. v. Hastings Corporation, 1950, 94 Sol. Jo., 225, C.A.).

There is no question as to the fact of malodour from fish processing (fish meal production). In fairness to the processor it should be stated that part of the odour stems from the raw material itself even before it is processed—and it is not always the salty tang which can be smelled for miles around. With fish, it is inevitable that within an hour after death and exposure in a suitable environment decomposition sets in. This is proved by the characteristic fluorescence of putrefaction which can be shown to exist 24 hours before it is ascertained by smell (227). In many cases, fish arrives at the fish meal plant in a more or less advanced stage of decomposition. Vapours from the processing of herrings are released to atmosphere only after treatment, but the treatment is not always effective. It consists of burning the smell, with all the well-known drawbacks of the method. Or, liquid agents are added to the load of herrings before processing, or blown as an aerosol into the effluent gases. Neither method is universally applied.

At a Symposium held at Doncaster Technical College (14th March, 1968), J. GRAHAM of the Torry Research Station reported a test at the Grimsby Fish Meal Company to establish the feasibility of odour suppression in the white fish meal industry (228). Products and byproducts in all stages pass through totally closed systems. Odour polluted air is not discharged to the open, but passed over a condenser and, after having cooled down, is recirculated to the drier. The condenser is of the indirect type, i.e., no contact is made between cooling water and polluted air. Only the condensate is polluted, and not the much larger volume of cooling water. The following notes are quoted from his paper:

(1) Recirculation of air can be recommended. In the trial, virtually 100% recirculation was achieved and no vapours were discharged to atmosphere.

(2) Existing plant with steam heated driers can be adapted to operate with a high degree of recirculation.

(3) It will be possible to operate newer types of steam heated driers with recirculation.

(4) Air circulation can be employed with scrubber condensers and indirect condensers.

(5) The cost of adaptation or installation will be low.

(6) The quality of the fish meal, the output of the drier, and working conditions in the factory will be unaffected.

The consumption of water by a spray or scrubber condenser with direct contact between cooling water (cf. Table 44) and polluted air is about 35 t/h or 1 t/h load of raw material. The dry-scrubber, where there is no direct contact, requires only about 1·4 t/h per ton of load for making-up, but will contribute only 1 t/h of condensate (waste effluent) per ton of load. The difference between 35 t of waste effluent going down the drains every hour for every ton of load, as against 1 t of waste effluent produced by the indirect condenser will decide for the latter. Not only is it the cost of water required by direct condensing which makes indirect condensing the method of choice, but that large volumes of heavily polluted water would move through the town sewers, sending up their evil smells through every vent shaft and manhole.

Rotary driers used in the production of fish meal are either directly heated from the furnace, Figure 9, or indirectly by steam flowing in the drier jacket, Figure 10. The steam drier is the most important single odour generator in the fish meal plant. Cookers and presses also emit large volumes of powerfully and obnoxiously smelling vapours. The principle of gaseous effluent deodorization is the same in either case: hot vapours

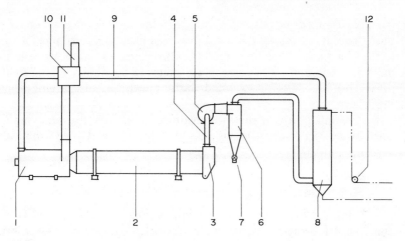

Figure 9   Deodorizing equipment for direct rotary driers. (*Courtesy Brødr. Hetland, Bryne, Norway*)

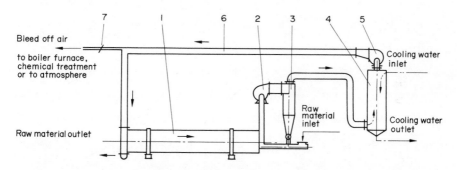

Figure 10   Deodorizing system for steam driers. (*Courtesy Myren, Oslo, Norway*)

The exhaust gas is discharged through the inlet of the drier (1) and blown by means of the fan (2) through the cyclone (3) into the scrubber (4).

The body of the scrubber is cylindrical, internally fitted with a number of horizontal perforated trays of special form and arrangement. Cooling water (sea- or fresh) enters at the top of the tower and falls by gravity from tray to tray forming a curtain of water. The gases flow in countercurrent movement thus being intimately mixed with the water. This causes condensation of water vapour and condensable constituents of the gas.

Before being returned to the drier (1) by means of another fan (5) and return air duct (6), the air passes through an entrainment separator for the removal of water droplets. A part of it is bled off and led to the furnace, chemical treatment, or released to the atmosphere.

The return duct is provided with a branch pipe and damper (7).

from the drying drum pass via a cyclone to the cooling tower and back to the furnace in the case of the direct drier, but back to the drum in the case of the steam drier. The direct driers oxidize the smell by combustion, the indirect driers absorb the osmogenic material by wet-scrubbing in the tower and dispose of the polluted washing water by discharging it into the drains.

The manufacturers of the equipment quote the cost for removing the smell by a direct rotary drier, as per ton of finished and dried fish meal,

Electrical energy	14 kWh
Fuel oil	9 kg
Maintenance	0·02 working hours (about 1 minute)

The indirect steam rotary drier requires only about one-third of the above electrical energy, but the price of supplying the cooling water is the determining factor.

21 Cooking smells

Great is the difficulty of designing satisfactory ventilation of a room which is full of food, but must not smell of any; which is close enough to the kitchen so that food served in this room is still sizzling hot, but must not harbour any of the hot kitchen fumes; and which produces sufficient air changes to satisfy these requirements without being draughty or heavy on the purse for capital investment for installations, and running cost. Much as the diner appreciates the aroma of good food on his plate, he is averse to smelling the food served at the next table.

The ventilating engineer will take into consideration all known principles of his profession plus any possible foibles of hypothetical visitors to the establishment.

There is no panaceic approach to the problem. One which offers an optimal solution requires two ventilating systems: one for the kitchen which may be considered the 'production' area, and one for the dining room of the hotel or restaurant, the 'environmental' air. The two require different methods of approach.

Cooking smells in general are not offensive. A plateful of cabbage or fish and chips may be enjoyable but the same food cooked by the ton is offensive if the cooking smells waft into offices and flats above a central kitchen, or in the direction of prevailing winds, filling the house with sulphurous or fatty stale odours. Cabbage contains thermolabile sulphur compounds, such as S-methyl-L-cysteine sulphoxide. This is a free amino acid, breaking down to dimethyl disulphide, $(CH_3)_2S_2$ (263). A similar smell, dimethyl sulphide, $(CH_3)_2S$, is produced by cooking meat.

The air from kitchens also contains some allyl, $-C_3H_5$, compounds, which have a distinct and persistent smell becoming quite offensive in quantity, for instance, $C_3H_5.S.S.C_3H_5$, which is diallyl disulphide, and allyl propyl disulphide, $C_3H_5.S.S.C_3H_7$, both from onion and garlic. There is also a considerable amount of moisture in kitchen air which requires destructive deodorization before it is released to atmosphere either at roof level or above, or near ground level. Air from the dining area requires no treatment at all. By separating the two systems it is possible to design each one for maximum effect.

Existing ventilation systems should be investigated for both function and effect before their incorporation in a new or a deodorization scheme is contemplated. The inner wall of ducts and, especially, bends will be found heavily coated with waste and grease and should be cleaned thoroughly, although this may be difficult. New ducting may be advisable in many cases. The pull of the fan will be found to be lacking not only because the motor is of an old type but, mostly, because it is coated with dust and grease, and the blades are heavy with filth. One cannot expect it to provide a steady airflow across hoods and other air intakes which are, mostly, of the large area type. With such equipment no control can be exercised of the random dispersion of fumes in ambient air. Unless white steam is given off the processed material there is little to indicate the flow pattern. Because of the invisibility of the matter polluting the air it is usually assumed that all is entering the ducts.

There is a simple way of making air currents in a kitchen visible. A few puffs of tobacco from a pipe blown gently from various directions across an intake will show how the air moves. Similar tests have been carried out in kitchens by placing a large pot of water on the range and bringing it to the boil. The pot is then placed on different sections of the range so that the steam from the boiling water rises from various positions relative to the hood extending over the range. The hood covers the area of the range quite adequately and will, in fact, be overhanging it by about 6–9 in on either side. With the water boiling gently, the steam will rise directly into the cowl when the pot is placed in the centre of the range. When moving it to the marginal areas the steam will be found to move off in practically any direction, mostly following the natural draught pattern. Only little, if any, of the steam will get into the hood and the ducted system. Once, however, the steam has risen to the edge of the cowl it will be drawn into the hood.

This picture changes abruptly when the water is boiling furiously and jets of steam are spouting from the pot. The thermal energy of the water molecules is then such that their movements occur at 'escape velocity', that is to say they leave the mass of molecules at high speed, thus getting out of control of the ventilation system. Efficiency is greatly impaired under these conditions.

Tests have been also carried out with vegetable boilers under their own individual hoods. On opening a fully loaded boiler the steam billowed out and up forming a mushroom cloud of which only the central core was taken into the collecting cowl. By the time that portion had been drawn into the duct the side lobes of the steam cloud had risen above the hood and were slowly dissolving under the ceiling by condensing on it and on sky lights and windows. The kitchen was filled with the typical sulphurous smell of hot moist cabbage in spite of the exhaust fan working at full power.

Failure of the fans to do their job properly is due mostly to two causes: the insufficient pulling power and the existence of cross draughts. If the prescribed number of air changes per hour is to be achieved and maintained, the room, be it a kitchen or entire factory, must be at a slight negative pressure all the time, so that any draughts are always towards, but never away from, the fan.

The number of small eating places in heavily populated sections of a city increases at an alarming rate. Alarming, because the preparation of food, disposal of kitchen waste, and control of cooking odours pose real problems of hygiene. Whether the little restaurants are on the ground floor or, for greater attraction and intimacy, in the cellar of an ordinary small house which was never designed for that purpose, nor can ever be properly adapted for it, the kitchen is nearly always underground and ventilation is either by the simple device of 'open window' or, if of advanced conception, by nine-inch fan mounted in the wall or a window pane.

The heat in such a kitchen must be experienced to be believed and so most of the personnel work in little more than a pair of trousers and vests or short-sleeved shirts. They sweat profusely. The food prepared in these kitchens tastes delightful when served at the table, but in the preparation of it heavily scented spices, and highly flavoured fish or meat are often used. In concentration, together with the hot and sometimes acrid fumes of cooking oil which may not be changed very often, pungent smells are produced which pervade the neighbourhood. 'Every person is entitled, as against his neighbour, to the comfortable and healthful enjoyment of the premises he occupies, whether for business or pleasure.' This sounds like mockery or utopia in the ears of people who live next to these restaurants, whether they prepare Bombay Duck (which is a fish dish) or other delicacies.

Deodorization under these circumstances will not, and often cannot, be contemplated by the proprietor, even in the face of a threatened closedown by the Local Authority, or by an injunction. Quite apart from the administrative aspect, there is the technical difficulty of dealing with heavily polluted hot air by any other method than exhaustion at the highest possible level, above the roof for preference. When this is suggested the Local Authority will instruct the owner that he must not run a duct up the front of the house, nor at the back, and that, whatever else he does, he must not cause any nuisance to other people, for instance, by blowing the extracted untreated air into the street either directly above the pavement, or through the top of his shop front, or through a light shaft where it would penetrate into all windows. The situation amounts in most instances to 'doing something without doing anything'.

There is, though, one way open at least to those with two (albeit rare) resources: extra capital for investment, and extra space to invest it in. The arrangement which has been found to work satisfactorily is depicted in Figure 11. A hood over the cooking equipment (range, grill) is the intake end of a ducted system which leads to ultra-violet deodorizers and into an air conditioner which cools the air. Immediately behind the hood, but of easy access, is a set of grease filters which must be cleaned daily by immersion in hot caustic soda or any of the commercially available degreasing agents. The duct should rise from the grease filter to the irradiation chamber, so that any fat which may have passed the filters will tend to run in the direction of the intake end, where a small grease trap, such as a hole about 2 cm in diameter, has a catch pot underneath it to collect it. If the air in the kitchen should rise above, say, 40°C (about 100°F) fresh external air should be induced before the hot air reaches the ultra-violet generators.

Behind the irradiation chamber is the pulling fan which draws the polluted air into the system propelling it towards and through the air conditioner which, in fact, need only be an air refrigerator. A bleed of external 'fresh' air may be provided just prior to the treated air entering the refrigerator section. If this might make the refrigerator too big, external air can be injected into the kitchen without any prior treatment at the rate of 10 % of the fan rating. A further simplification may be achieved by using a low-speed fan so that the air moves but slowly through the duct, whose dimensions should be calculated to give 5–6 sec dwell time. The irradiation chamber can be omitted and the ultra-violet generators are then mounted direct on the duct wall, the duct now acting as a long narrow chamber.

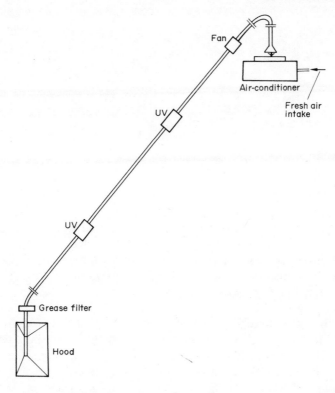

Figure 11   An arrangement of ultra-violet tubes for the destructive deodorization of air working in conjunction with an air conditioner in a restaurant kitchen which has no outlet for the polluted air.

22 Catering industry

Meat served in grill rooms and large catering establishments is often treated with a tenderizer (pre-digested meat treated with pepsin in acid solution) or soaked in oils specially prepared to achieve a similar effect. A meat tenderizer of natural composition is papayotin (papain) a vegetable enzyme, from the dried milky juice of the papaw, the edible fruit of the papaya tree (*Carica papaya*, South America).

On putting a piece of meat over the fire the oil in which it has been soaked burns up and the smell of acrolein arises. Most vegetable and animal fats have melting temperatures below 50°C, and boiling temperatures below 80–100°C, so that overheating, i.e., burning, takes place almost the moment the meat is put on or over the fire. The blue smoke curling up is finely divided solid matter, i.e., carbon particles.

The dairy industry produces an effluent having a biological oxygen demand value of 150 to 500, equivalent to that for the sewage treatment for 1000 to 3000 people. With the exception of cheese products, there are none which would pollute the air to any extent. The flavour of cheese is mostly caused by the breakdown of protein containing about 0.8% sulphur. In the process amino acids are degraded to amines and acids, the former being responsible for the odour. Another substance which gives cheese its particular flavour is acetylmethyl carbinol, $CH_3.CO.CHOH.CH_3$. The average content in cheese is 3.25 mg/100 g, but heavily flavoured cheeses, such as Stilton, Gorgonzola, and others, may contain 8 mg/100 g. Mild cheese, on the other hand, is flavoured by 0.1 mg/100 g, or less. Acetylmethyl carbinol when oxidized, yields diacetyl, $CH_3.CO.CO.CH_3$ which gives butter its fresh taste. Cheese may also contain β-methylmercaptopropionaldehyde which is an important element in its characteristic flavour; in concentration it smells of raw pumpkin. Milk beginning to go off also smells of this compound.

Coffee beans consist of a variety of substances which in the green bean do not become apparent to the senses of taste or smell; it is upon roasting that great changes occur. Chaff and other refuse must be cleaned from the beans; caffeine, the most valuable ingredient of the coffee bean, may decompose on the beans being heated too much, and spoil the flavour—acrolein is probably formed and is one of the causes of the strong and objectionable odour from coffee roasting establishments.

An analysis of coffee beans gives

Acetaldehyde	$CH_3.CHO$
Diacetyl	$CH_3.CO.CO.CH_3$
Furan	$CH.CH.CH.CH.O$
Furfural (Furfuraldehyde)	$C_4H_3O.CHO$
Furfuryl alcohol	$C_5H_5O.CH_2OH$
Furfuryl mercaptan	$C_5H_5O.CH_3S$
Hydrogen sulphide	H_2S
Methyl-ethyl carbinol	$CH_3.CH_2OH.CH_2OH.C_2H_5$
Pyridine	$CH.CH.CH.CH.CH.N$

In concentration, any and all of these are highly offensive. That concentration may well occur in practice is shown in the 1965 Emission Inventory for Chicago (159) which lists 55 machines roasting a daily total of 175 tons of green coffee beans within a relatively small area of 9 square miles. The area is densely populated. The total air pollution from the premises of these 15 processes is some 50 t of osmogenic particulate matter. There is dust and refuse from the cleaning stage, odour and blue fumes (carbonized particulates) from the roasting stage, and chaff and other waste discarded at the final inspection.

23 Fish friers

The frying of fish produces what are pleasant and appetite-stimulating odours to one person, offensive smells of hot fat with a trace of amines and vinegar to another. Local Authorities usually ask for some means to be taken by the proprietor of a fish and chip shop to prevent the direct discharge into the street of fumes collected from over the frying vessel. The usually recommended, and usually fitted, grease trap in the duct leading from the collecting hood to an outlet at or above street level is more a show of willingness to comply with regulations or recommendations, than a suitable means for reducing an odour. At best, some of the hot oil is retained in the trap which, if not cleaned out daily, is soon choking the air flow.

Various methods of air deodorization have been tried, but none have quite succeeded. The shop owner cannot spend relatively large sums of money for the abatement of a hypothetical nuisance which, to him and his customers, is a pleasant and looked for (or sniffed for) advertisement. Forced ventilation using charcoal (activated carbon) or other filters are useless as they are not maintained in working condition and become hopelessly choked with fat within a few days' operation. Destructive deodorization by ultra-violet irradiation of the air suffers from the disadvantage that an irradiation chamber is required and that the hot fat in the air passing the units will soon coat them with an opaque layer. Such installations, described in relation to cooking smells generally, are unfortunately too expensive by far for most of the shops. It is a farcical situation that a fish frier's shop—such a familiar and widely appreciated institution—cannot be rendered entirely inoffensive within its financial framework.

24 Merchant shipping

Under this heading must be considered the transport of a variety of food-stuffs in refrigerated cargo ships or vessels storing their load in closely controlled atmospheres. Fruit from overseas must be harvested and shipped while still unripe. During the voyage, it is allowed to reach a more advanced state, but to mature fully it is kept in cold store until a short time before sale.

A fruit beginning to mature gives off the gas ethylene $CH_2.CH_2$ as a metabolic waste product. When immature fruit, for instance apples and bananas, are immersed in an atmosphere carrying a little ethylene, this will act as a strong stimulant furthering the maturing process. Once on board ship, even a single fruit in the state of giving off ethylene as a respiratory product can set off a whole hold of immature fruit on the process of nearing maturity at an untimely moment. This would cause the shipper considerable loss.

The only way to control the 'breathing' of fruit, i.e., the emission of ethylene, is by adding ozone to the circulating air. Ozone will break down ethylene, $CH_2.CH_2 + 2O_3 \rightarrow 2CO_2 + 2H_2O$, thus not only removing the stimulant (ethylene), but also adding a respiratory process yielding the products of complete combustion, water and carbon dioxide (276). The ozone concentration in air suitable as a control of the ripening process, varies within relatively large limits. Apples, peaches, and bananas should receive ozone in concentrations not exceeding 2 ppm. More will be deleterious and oxidize the tissues. But oranges and citrus fruit can take up to 40 ppm of ozone (277). The presence of ozone in the air of cargo ships will not only delay very considerably the maturing process, it will also prevent the onset of stem rot in hands of bananas. DR GANE of the then Department of Scientific and Industrial Research (276) (later a Division of the Ministry of Technology) gives the following data referring to bananas after 18 days' storage at 12°C (55°F).

Effect of ozone in controlled atmosphere

	$10\% CO_2$ +low O_3	$10\% CO_2$ +high O_3	$10\% CO_2$ no addition	
Length of rotted stem in inch (cm) at the proximal end	1·5 (3·8)	0·6 (1·5)	3 (7·5)	2 (5·0)
at the distal end	0·8 (2·0)	0·4 (1·0)	1·25 (3)	1 (2·5)

The output of ozone as produced by the ultra-violet irradiation of air rises in direct proportion to humidity, a condition which the atmosphere on board ship or in cold storage places requires. Cold dries out the tissues whether vegetable or animal, and to prevent shrinkage and loss in weight various degrees of relative humidity are specified for various commodities. Moisture in air is also a cause of mould formation; but the microbicidal effect of ozone is allowed to control the situation. Air carrying between 0·5 and 1·0 ppm ozone is allowed to flow for 3 hours daily into every nook and cranny between fruit boxes, and the mould is soon eradicated.

Meat presents different problems. The carcases hang closely packed together in the holds. They often fall victim to the attack by *Mucor Aspergillus niger, Thamnidium* and *Sporetrichum carnis,* forming filaments of mould. So infected meat is retrieved by cutting out the infected pieces. This is a loss of substance which no importer likes to see in his ships. Ultra-violet radiation is ineffective because spatial arrangements are such that generators cannot be brought into position to irradiate the meat effectively, as all space is taken up in hanging the carcases. Moreover, direct ultra-violet irradiation of meat would have a deleterious effect. Haemoglobin, a chromoprotein and the colouring matter of the erythrocytes (red blood corpuscles), changes to methaemoglobin, a brownish substance, under the impact of ultra-violet rays, thus discolouring the meat. This is a direct consequence of nascent oxygen and ozone oxidizing the haemoglobin. The spectral lines of the generator between 2100 Å and 3000 Å are the most active in this: see Table 23.

Natural light (sky light or sunshine) has much less effect on the colour of meat exposed to it. At 3132 Å the effect is only 2%. Fat under the action of ultra-violet rays, that is in the presence of nascent oxygen and ozone, will oxidize, i.e., turn rancid. Oxidation of fats in a dry atmosphere is a relatively slow process which can be accelerated by a rise in temperature, and which will produce saturated fatty acids (general type $C_2H_{2n}O_2$).

Table 23

Formation of methaemoglobin by ultra-violet radiation

Wavelength Å	% Effect
2270	50
2400	90
2537	100
2650	51
3000	18

These do not produce a smell if $n > 15$. In a hot, yet moist, atmosphere low-molecular weight fatty acids with $n < 15$ are mainly produced. These have a very objectionable smell. Hydrolysing enzymes (lipases) can cause deterioration at temperatures as low as 30°C. The general formula for the oxidation of a fatty acid is: acid $+ O_2 \rightarrow$ ketone $+ CO_2 + H_2O$.

$$C_nH_{2n}O_2 = CH_3(CH_2)_{n-2}.COOH + 2O \rightarrow$$
saturated fatty acid

$$CH_3.CO.C_{n-3}H_{2n-5} + H_2O + CO_2$$
ketones

$$C_nH_{2n}O_2 \qquad + \qquad 2O \qquad \rightarrow$$
saturated fatty acid + atoms of oxygen

$$C_{n-1}H_{2n-2}O \qquad + \qquad H_2O \qquad + \qquad CO_2$$
ketone   +   water   + carbon dioxide

The only safe method of dealing with meat mould is to fill the holds with ozonized air. Ozone plus carbon dioxide (10 %) is the most useful combination of value to meat shippers (278).

It is a widely-held misconception that microbial life is extinguished in refrigerated space. This is true only for the deep freeze boxes. Many micro-organisms perish at temperatures well below blood temperature (37°C = 98·6°F), but then, most of these are found in warm-blooded animals including man. But there are a significant number either harmless or pathogenic to man, which remain quite active at temperatures just above—and some even below—freezing point, 0°C = 32°F. This group of bacteria is known as psychrophils (Gk: *psychros*, cold; *philein*, to love) and confronts the refrigeration industry with problems, as these microbes are much more common than was once thought. They are found in the intestines of fish, on chilled or frozen meat, in raw or pasteurized milk, butter, cream, in tap water and ditches and in garden soil, on vegetables whether frozen or fresh, in flour and, as one might expect, in air.

Psychrophils are affected by higher temperature. Taking one of them, *Pseudomonas fluorescens*, cell count and temperature relate an interesting story.

Temperature	20°C	5°C	0°C	−3°C
Cell count	995	1850	1590	1200

Pseudomonas exist in various strains which spoil certain foods, mostly dairy products, by producing unacceptable off-odours and off-flavours, sometimes also pigments, in butter, cheese, and other products; also in milk unless it is stored at low temperatures. Other *pseudomonas* prefer meat, for instance frozen lamb and pork, and chilled poultry. They do not like beef which is more often spoiled by micrococci, another member of the psychrophilic group.

The bacteria most dreaded by man in connection with the storage and spoilage of food, even at temperatures as low as 1°C, are the *salmonellas* which proliferate extensively at this temperature near freezing point. This makes the domestic refrigerator quite unsafe from them as it maintains a temperature between 3° and 5°C. Food is not safe from *salmonella* whether it has been put in the refrigerator free from it, or already infected. Food transferred from a deep-freeze (−10° to −15°C) to a domestic refrigerator is an excellent growing medium for the *salmonellas* and other micro-organisms.

Chromogenic bacteria also belong to this group and are a particular nuisance in that they produce pigments in the course of their normal metabolism. They grow preferentially on chilled poultry and cause the well-known discoloration of the birds. Other psychrophils are *serratia*, *achromobacter*, *flavobacterium*, and others. All, or most of them, produce indole as a metabolic produce, and decompose protein. Either action is accompanied by distinct and equally repulsive smells. Both types have low threshold perception which, from our point of view, is a useful thing as it helps us to detect a piece of poultry going off long before the protein decomposition products have reached toxic level.

Antibiotics have proved ineffective in the fight against psychrophils because they have developed antibiotic-resistant strains. The only safe remedy is bactericidal radiation (2537 Å), but this will destroy bacteria only on the surface of the food; those within deeper tissues are safely shielded. The bactericidal effect of ultra-violet rays is explained by the fact that nucleic acid, the building stone of bacteria and viruses, is denatured maximally upon absorbing radiation at 2537 Å. Bacterial nucleioprotein and RNA (ribose nucleic acid) also absorb the radiation heavily.

Shipping economy demands the use of the holds, whether refrigerated or not, for any kind of goods without tainting one load with the smell of that carried in the same hold on a previous journey. (Though one would not normally store butter in the same hold of a ship which had carried onions, or pears, or apples, or bananas.) If, after every journey, the whole cargo space is flushed with heavily ozonized air, not only will the holds and any permanent structures in them, be deodorized and cross-tainting avoided, but also vermin, especially of the kind mostly moving on the floor, will be exterminated by breathing ozone at lethal concentration.

It is not often that ultra-violet generators are mounted individually in holds. A preferred technique is to fit a number of generators in the fan chamber and have the ozone mix with the air and circulate throughout the cargo space. Internationally recognized concentrations of ozone in air, are

$$6 \text{ mg}/1000 \text{ ft}^3 \text{ (35m}^3\text{) for half-fruit standard}$$
$$12 \text{ mg}/1000 \text{ ft}^3 \qquad \text{for full-fruit standard}$$

These standards refer to the volume of refrigerated holds and the degree of ripeness of fruit. The electrical control gear of the ultra-violet generators can be arranged to produce three levels of ouput, i.e., 6 mg, 12 mg, and full unrestricted output for scavenging. Assuming a generator which will produce 300 mg of ozone per hour, and a hold of 6000 ft^3 (200 m^3) the requirements are: (1) an output of 36 mg/h, (2) an output of 72 mg/h, and (3) scavenging (full) output of 300 mg/h.

25 Maggot breeding

A maggot is a wormlike larva of dipterian (two-winged) flies, especially of the green blow fly, *Lucilia sericata* and of *Phormia regina*. In such holometabolous or endopterygotous insects, the larva develops into an adult by sudden metamorphosis. The larva (maggot) differs fundamentally from the adult form (fly). A maggot has no appendages or well marked head, and feeds on decaying flesh by secreting allantoin, a diureide of glyoxylic acid,

$$NH_2.CO.NH.CH \begin{array}{c} CO\!-\!NH \\ | \\ NH\!-\!CO \end{array}$$

Allantoin hastens the formation of epithelium (covering layer of the skin) and has been employed to clear wounds and ulcers of skin debris, especially in the treatment of osteomyelitis.

Maggot breeding establishments are mostly concerned with supplying anglers; a few also provide maggots for medical research.

The process is simple enough. The raw material is fish or offal from abattoirs which is allowed to become fly-blown. The larvae are then permitted to feed on the fish or meat, and are collected for sale as soon as they have reached a respectable size, but before they are due to change into a chrysalis. In this third (penultimate or pupa) stage in the development of holometabolous insects locomotion and feeding stops, but great developmental changes occur. Because fish are attracted by the wriggling larvae, it is important to collect and deliver the maggots in good time before the onset of the final metamorphosis.

The parentage of the larva is all-important. The bigger the maggots, the higher their price. The really professional breeder—there are not many of them—will import special breeds of African giant flies which he then keeps in a 'nuptial chamber', guarded by a double fine wire netting and accessible only through one, often two, air locks. This is to prevent any common (European) bluebottle from getting in and spoiling the breed.

Fish and meat is brought into the chamber for the flies to lay their eggs on, and after a few hours is taken outside and put into bins about 3 ft × 6 ft × 1 ft 6 in high, sprinkled over lightly with sawdust, and covered with sacks or pieces of hessian so that the maggots will keep warm, yet have air to breathe. Neither fish nor meat deteriorate under the maggots, because their allantoin turns the surface of the carcass into a mummylike condition. The maggots feed voraciously and grow quickly to the required size which, for African maggots, is up to 1 in long.

If fresh fish or meat is used, no smell will develop, but if the feeding stuff is old and already entering the stage of putrefaction before it is fly-blown, it becomes a strong source of odour. During their larval stage the maggots produce large quantities of trimethylamine, $(CH_3)_3N$, which smells of ammonia when highly concentrated: workers in a maggot-breeding place always refer to it as ammonia. If, on entering a maggotorium, the smell is of rotting fish, the metabolism of the maggots is out of order: trimethylamine in low concentrations smells of rotten fish, not of ammonia.

There is an important point here in connection with the breeding of maggots on a large scale. During the period of feeding a strong lacrimating smell rises from the maggots, penetrating the whole building. Maggot breeders are inclined to diagnose the offensive odour as 'ammonia', introducing hydrochloric acid into the wet-scrubbing process in the hope of controlling the nuisance. There can be no chemical reaction to suit their purpose since it is trimethylamine in concentration—and not ammonia—which causes the smell and the irritation of the eyes. Since there is no reaction between hydrochloric acid and trimethylamine, the smell is not abated. So more hydrochloric acid is added to the fumes until a 'clean', i.e., slightly acid smell, results. This is caused by droplets of un-reacted acid which will corrode materials and damage, even kill, vegetation around the maggotorium. Pulmonary damage to man and beast is inevitable under these conditions.

There may be two sources of smell in a maggot breeding establishment. If the breeding stuff is brought in daily and fresh or kept in refrigerated space, there is only the smell of trimethylamine in the breeding house. If the place is neglected and rotting raw material is left piled high in the open or arrives in a deteriorated condition, the smell contains hydrogen sulphide in addition to trimethylamine and is an uncontrollable source of the most offensive odour; such a condition contravenes the bylaws of any Local Authority.

Polluted air from breeding houses—the air from the nuptial chambers is never treated, as it has no bad smell of its own—can be treated successfully only with nascent oxygen. A ducted ventilation system is required which draws the air into an expansion (irradiation) chamber. Suspending ultra-violet generators in the maggotorium above the bins is not good practice: it would be impossible to treat all the air in the place, would require a large number of generators, and would be expensive. Moreover, the rays will kill the maggots which have no protective pigments. Breeders dye their maggots occasionally just prior to sale, because fishermen seem to think that coloured maggots are a greater lure. Whatever merit there may be in this assumption, dyed maggots are just as sensitive to ultra-violet radiation, but neither ozone nor nascent oxygen has an ill-effect.

26 Sewage and domestic refuse

As both sewage and domestic refuse are a consequence of habitation it is proposed to deal with them under the same heading. The amount of these wastes is surely a true indicator of the trend in which the development of a population is moving. The one increases with the number of people, the other with the necessity of canning and preserving food in times of plenty, and with the misconceived application of technical genius to producing prepared foods which, for reasons of safe-keeping and greater sales appeal, must be packaged. Not only has the value of the materials used in the packaging industry risen manyfold during the last twenty years —only to end on the waste grounds and in municipal incinerators—but the sheer bulk of these wasted materials and products threatens to overflow all amenities provided for and available to dealing with domestic refuse, Figure 12. The weight of household refuse in England has trebled in the last thirty years (150).

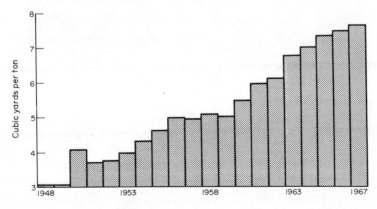

Figure 12 The changing density of refuse in a typical British city. (*From* J. E. Busfield, *Surveyor and Municipal Engineer*, 3.12.66)

137

Improvements are frustrated by a vicious circle. It takes years to plan a new waste processing plant; years to get permission, and the necessary funds, to build it; and all the time the populace increases. By the time all the official proceedings have been through their appropriate stages and sanction has been given, it is found that the new works, if executed according to approved plans, would again be inadequate within a few years. There is neither space, nor the money available, to build today for a time 20, perhaps 50 years, ahead.

This is perhaps the most important, the most urgent reason why new methods of dealing with human waste and scrap are developed at speed, some proving inadequate, others standing their ground. Promising above all are methods of converting, not destroying, the waste material.

Figure 13   The compost house of Bangkok's refuse composting plant, handling 300 t/day crude refuse. (*Courtesy John Thompson Compost Plant Ltd.*)

Apart from metal tins and cans, and broken glass in domestic refuse, most of it is of an organic nature suited, in combination with sewage, for conversion to fertilizer.

Rendered, dried, and ground waste of household food materials is known as garbage tankage and contains $N-2\cdot5-3\cdot5\%$; $P_2O_5-2\cdot5\%$; $K_2O-0\cdot5-1\cdot0\%$. The principle involves no unsolved problems. Composting is well known to every farmer, but the country smells, quite inoffensive in their appropriate setting, are repulsive to townspeople, and the processing of sewage and refuse near a densely populated town will invariably cause the people to see flies and worms, and perceive nauseating odours everywhere, even if there are none. It is therefore an essential point to watch that any such plant is completely mechanized with as small a labour force as possible; that the whole plant is completely enclosed to work as an odourless factory; and that the dry domestic organic refuse and dust is used to take up the sewage sludge.

The first waste conversion plant of that type was built by the States of Jersey Sewerage Board in the British Channel Islands to the design of S. A. GOTHARD (149). It is of particular interest in that it not only relieves a difficult situation arising during the holidays when the number of indigenous population is nearly doubled by the influx of visitors, but also produces a saleable product, i.e., compost of the highest quality.

The flow diagram (Figure 14) gives a clear indication of the process. Domestic refuse is collected in the usual way and enters at the works first a separation section where non-compostable matter is picked by hand, dust is transferred to an automatic weigher and then stored, and ferrous metals are sorted by a magnetic drum, salvaged, and sold. Abattoir and rural wastes are then broken down mechanically and mixed with dust from the storage bins, and thickened sludge from the clarifiers is added in automatically controlled proportion. This mix is transferred to fermentation cells and, finally, to the stores for maturing.

When the plant was first operated, a maturing period of three months was required. Not only did this cause storage problems, but the odour rising from the fairly large amount of compost was strong and objectionable. Qualitative analysis of the gases showed the presence of hydrogen sulphide, amines, mercaptans, ammonia, and other less or non-osmogenic constituents. A reduction of the moisture content of the compost improved the oxygen deficiency in the fermentation cells which had caused the generation of sulphydryls. The required aerobic fermentation is achieved in a battery of 36 cells arranged on six floors. The mixture is loaded into the cells on the top floor (Figure 15). After 24 hours the bottom of these cells opens and discharges the already fermenting waste into the cells on the floor below, and so forth until on the seventh day the cells on the bottom floor discharge the ready compost into vehicles (Figure 16). This method of fermentation by stages not only turns and aerates the matter thoroughly and automatically once every day, but also leaves the topmost cells free every day to receive a new charge. At about midweek, or before, the temperature in each cell has reached 75°C which is sufficient to kill all pathogenic microbes, seeds, and so forth. This is

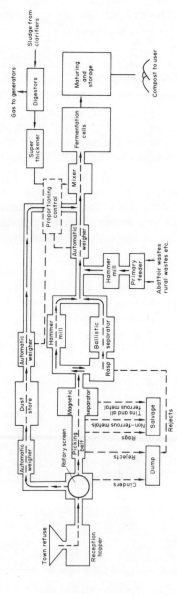

Figure 14 Flow diagram of the Jersey, C.I., refuse treatment plant. (*Courtesy States of Jersey, C.I., Sewerage Board*)

Figure 15   Loading cell, Jersey, C.I. (*Courtesy States of Jersey, C.I., Sewerage Board*)

Figure 16   Fermentation building, Jersey, C.I. (*Courtesy States of Jersey, C.I., Sewerage Board*)

several degrees above normal pasteurizing temperature. The finished compost is absolutely sterile throughout.

The power requirements for turning some 480 t of fermenting waste, which is the capacity of the fermenting house, is less than 2 kW.

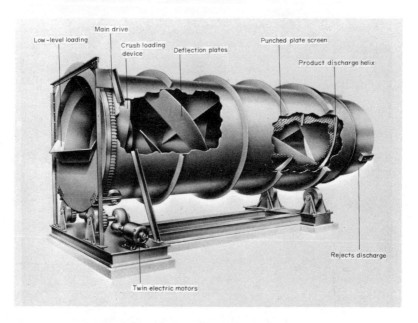

Figure 17   A Vickers seerdrum. (*Courtesy Vickers Ltd.*)

The sewage treatment uses primary sedimentation, tapered aeration, and final settlement. The sludge is digested and thickened. The dewatering of the sludge caused some difficulty, economic more than technical, because vacuum filtration, the best method, is expensive, but essential. Sludge gas contains only small proportions of hydrogen sulphide, but up to 75 % of its volume is methane. It is a valuable and convenient fuel and sufficient is produced to provide power for all machinery in the plant.

Table 24   *Average composition of sludge gas*

Methane	60–75 %
Carbon dioxide	15–30
Nitrogen	1–10
Hydrogen sulphide ⎫	
Hydrogen peroxide ⎬	> 1
Other gases ⎭	

As long as the volume of hydrogen sulphide remains below 1 % it does not constitute an odour nuisance in sludge gas.

Although there is practically no smell from the plant working on a continuous seven-days-a-week basis, the introduction of a 5-day working week with a 48 h break has given rise to a rather serious smell problem. The installation of a wet-scrubber using water sprays has practically eliminated the nuisance. A similar works has been built in Cape Town (186).

A step further in dealing with refuse, especially of the domestic type, is to compact it at source, i.e., in each house, then collect these neat little bundles and dispose of them in a suitable way at a place well away from town. A small, yet highly efficient compacting machine has been built in the United States to avoid in this way the pollution from many small incinerators.

However, not every sewage works has been built recently. In fact, there are quite a number which are 50 years old, and some are even older. Not only are systems of operation and machinery lamentably inadequate, but the capacity of these older works is heavily overloaded because of the increase in population. The inadequacy is further underlined by the layout of the treatment plant which, years ago, favoured open channels and un-covered tanks. These, because of insufficient aeration, produce strong smells of sulphydryls and hydrogen sulphide in addition to the natural smell of sewage (indole and skatole). To deal with such a problem by any method of deodorization, whether destruction or masking, is highly un-economic. This is due to the lack of enclosures for the polluted air, its large volume, and the relatively low concentration of the pollutants. The remedy lies only in the rebuilding of the works on modern lines and with sufficient capacity, having an eye on the steadily rising trend of population. This remedy, however, costs money, much money, from £3 to £12 per head of population served, or even more and is, really, an investment by the present generation for the benefit of the next one. Understandably, many a Council feels that however philanthropic such a suggestion might be, the Treasurer's report would draw the blue pencil across it.

There remains the middle path. One or two new tanks added will increase capacity; the main sources of smell should be covered up, and air flow in the works controlled for anti-pollution treatment in those sections of the structure which lend themselves to such treatment. Whilst these measures will by no means produce an ideal situation, they will offer an amelioration of conditions as reasonable as may be expected in relation to capital invested.

Anaerobic conditions in septic tanks, stagnation of flow in old sewerage systems, or interference by industrial waste effluents which are discharged into public sewers without proper treatment, are some of the causes of smells emanating from manholes in the streets. Sludge drying beds in older works, even when correctly managed, will release bubbles of foul gases when the sludge is lifted for transfer to other beds or removal. Raw sewage flowing quietly in an open channel need not be a source of offence in every case except when the sewage is septic. In rural areas, where the sewage collecting tank waggon may pump septic sewage from a house, a smell nuisance will arise at the works.

Normal raw sewage will give off fumes of considerable nuisance value at the point of measurement, i.e., at the flume, or at any other point of turbulence. Sludge presses, whether ancient or modern, are of necessity the origin of foul odours and, often, the cause of rather widespread osmogenic pollution when the expressed liquid is directed straight into the system.

Fermentation in sewage may be intended—as in composting sewage treatment—or unintended when micro-organisms are allowed to act upon the organic matter carried in sewage. Bacterial action is exothermic, and the energy is a biological requirement in the life of these bacteria. Biochemical oxidation occurs mostly in a watercourse when bacteria utilize the atmospheric oxygen which has dissolved in the water. Sources of sulphur, from which hydrogen sulphide can be formed by chemical or biological, i.e., bacterial, processes are industrial waste effluents, but even in their absence the danger of H_2S formation is always present owing to the relatively high content of sulphates, $-SO_4$, in faeces. Per head and day the sewers carry about 2·6 g of sulphates to the sewage treatment plant. This is for each group of 100 000 people a total of over 0·25 t of sulphate. Hydrogen sulphide is lethal at relatively low concentrations. Maximum acceptable concentration (MAC) is 10 ppm (15 mg/m^3) and 0·2% v/v causes death within a few minutes. In low lying parts of man-sized sewage pipes, over sewage or sedimentation tanks, and at other locations where maintenance men may have to work, the accumulation of hydrogen sulphide in air may often be lethal owing to it being heavier than air (s.g. = 1·190).

A visual indication of hydrogen sulphide in air is afforded by moist lead acetate paper. In the presence of H_2S the nearly colourless paper will turn black (283, 284). Concentrations up to 0·4 % have been measured. Accidents due to sudden immersion in gaseous hydrogen sulphide are possible, because at higher concentrations the gas loses its smell of rotten eggs and changes to a rather pleasant, though lethal, smell (151). This effect is, of course, not characteristic of hydrogen sulphide—which is the colloquial way of putting it—but of the human sense of smell. The biological effects of H_2S concentrations are listed for guidance.

Table 25 *Biological effects of hydrogen sulphide*

Olfactory threshold	$1·3 \times 10^{-3}$ ppm
Maximum acceptable concentration: no danger after prolonged inhalation	10
Slight damage after a few hours of inhalation	50–150
Maximum concentration which causes no damage after 60 minutes of inhalation	170–250
Dangerous effects after 30 minutes	300–600
Acute poisoning	600–1000
Death within a few minutes	2000

Hydrogen sulphide is highly soluble in water and dissolves readily in sewage. Normal H_2S concentrations are up to about 15 mg/l but in heavily contaminated (septic) waste it may be up to 50 mg/l. Sewer air not only has smells peculiar to the location, and is dangerous for reasons discussed above, but becomes the more dangerous when it is considered that the oxygen content of the liquid is reduced, particularly when the pipes are running to capacity. The decomposition process within the faecal mass in sewers is aerobic from the start, but when all oxygen absorbed in the water has been consumed, the process turns anaerobic, adding to the intensity of the smell. The absence of oxygen from that particular stream of sewage will cause oxygen from another stream flowing in a larger pipe, and into which the oxygen-depleted sewage carries, to be absorbed so that the air overlying the larger stream will become poorer in oxygen and constitute an asphyxiation hazard to the worker. After several hours the air in a sewer may have lost all oxygen and now consists mostly of nitrogen and hydrogen sulphide.

The amount of oxygen dissolved in sewage whether flowing or stationary, can be measured by means of a fully automatic meter giving readings from 0–15 ppm. The instrument has been designed at the Water Pollution Research Laboratory, Stevenage, Hertfordshire, England (a Government establishment), and is described by BRIGGS *et al.* (152).

Stagnation of the flow of sewage has been shown as a source of hydrogen sulphide. POMEROY and BOWLUS (153) have calculated the limits of velocity and temperature above which no hydrogen sulphide can evolve. They start from considerations of the biochemical oxygen demand for which the formula

$$BOD_t = BOD_{20} \times 1 \cdot 07^{(t-20)}$$

is given where t = temperature °C of the sewage, BOD_t = BOD of the sewage at temperature t and BOD_{20} = BOD at standard temperature 20°C.

The term BOD_t is called the effective biochemical oxygen demand and is also written BOD_{eff}. The compilation of effective BOD values is taken from a paper by MÜLLER (154).

Table 26   *Effective biochemical oxygen demand and temperature*

$t°C$	BOD_t/BOD_{20}	$t°C$	BOD_t/BOD_{20}
11	0·52	21	1·07
12	0·56	22	1·14
13	0·60	23	1·22
14	0·65	24	1·31
15	0·69	25	1·40
16	0·75	26	1·50
17	0·80	27	1·61
18	0·87	28	1·72
19	0·93	29	1·84
20	1·00	30	1·97

A temperature rise of 10° above standard doubles the biochemical oxygen demand (at 30°C).

Table 26a   *Minimum velocity and
effective biochemical oxygen demand*

BOD$_t$ mg/l	Minimum velocity of sewage	
	m/sec	ft/sec
55	0·30	1'0"
125	0·45	1'6"
225	0·60	2'0"
350	0·75	2'6"
500	0·90	3'0"
690	1·05	3'6"
900	1·20	4'0"

Sewage treatment works often incorporate the use of lagoons which, in fact, are huge settling tanks or storage tanks set into the soil, occasionally brick or concrete lined. The liquid in a lagoon is left for days without disturbing it. Anaerobic conditions are likely to develop at the bottom, rapidly spreading through the water. This becomes painfully obvious when the settled sludge is drained off and hydrogen sulphide bubbles, contained in the thick mud, burst open. There is little that can be done to cure or ameliorate this nuisance unless the lagoons are covered in and a deodorization system is operated.

It is accepted practice to discharge liquid residues from sewage treatment works into rivers. This has caused river pollution, but since the introduction of the River (Prevention of Pollution) Act, 1961, which is explained in the Ministry of Housing and Local Government leaflet No 39/61, dated 30th August, 1961, the River Boards have authority to stipulate conditions which will make any municipal or industrial waste effluent acceptable for discharge into a river. The Act itself, and all regulations issued by the River Boards, are only concerned with the sanitary aspect of rivers. If there is any process carried out on land which is likely to pollute a nearby river, whether by soiling the land between the factory and river, or by discharging effluents untreated, or insufficiently treated, into a river, the Board can apply to the County Court for an order prohibiting the use of the land for the purpose.

The severity of the 'available oxygen' problem in natural waterways is made clear by the statement that, by 1980, the biochemical oxygen demand of sewage and other waterborne waste in the United States will have increased so much that it will consume, in dry weather, all the oxygen carried, i.e., available, in the 22 river systems in the States (185). To safeguard river life, ELDRIDGE (155) quotes the following figures as minimum oxygen requirements:

 2 ppm for preventing anaerobic decomposition
 4 ppm for bottom-feeding fish
 10 ppm for surface-feeding fish

Most towns situated near the coast have, in the past, simply discharged directly into the sea. Depending on tidal flow patterns there was, in some cases, a need for passing the sewage through a detritor before discharging it. Depending on the formation of the coast-line, the population, and other relevant considerations, the outfall may have been directly at the foreshore, or taken by pipe a few hundred feet into the sea. The location of such a submarine outlet is given away by the seagulls floating on the water waiting for titbits to come up.

The fear of polluted beaches and adverse reaction by visitors to seaside towns, whether they are actually bathing and swimming in the sea or enjoying walks on the promenade, has put a stop to this rather crude, yet economical, practice, and more sophisticated schemes, like the one in Jersey, are taking its place.

Investigations into the bacterial contents of sea water along the coast line, and especially in some bays, of the Mediterranean between Spain and Yugoslavia, have shown that there are at least 20 000 $B.coli$/cm^3 sea water. This exceeds the limit of contamination set by the World Health Organization by a factor of seven. Another aspect of this is the increase in the growth rate and multiplication of algae which is promoted by the presence of $B.coli$. Resulting from this is an increase in phosphates and nitrates on such a scale that the water has lost its faculty of biological self-cleaning. Hence, even long waste pipelines leading into the sea, an idea which has been promoted by some people, will not solve the difficulties.

It has been found that using pipelines some 3–4 km (2 miles) long and releasing treated sewage only at outgoing tides, the tendency can still exist for the sewage to hang about the estuary or where the sea is buffeted between neighbouring or rocky coastlines, or islands. This was, in fact, revealed by investigations into sea currents around the shores of Jersey, C.I., when it was found that there was no point which could be regarded as fully satisfactory in this respect. And this with the nearest island, the Isle of Sark, being 15 miles (23 km), and the nearest point on the Continent being some 20 miles (35 km) distant. Even where the formation of the sea floor within, say, 5-6 miles (about 10 km) of the coast would permit the laying and anchoring of a sewage outlet pipe at not too great an expense, the hydraulic pressure would prevent the natural outflow of the sewage, and powerful pumps together with submarine valves would be required for discharging the effluent.

A new technique is now under consideration which will ship the sewage out to sea and discharge it at locations deep enough to make sure that no currents will bring it back to shore. This may entail journeys of 50-150 miles (100-300 km) out to sea. It is envisaged that such a journey would take place once or twice weekly depending on the tonnage of the vessel, and the quantity of sewage. The advantage of this technique, apart from the safe disposal of the effluent, is in that several towns along the coast can budget together and share the services and time of the ship. At some secluded spot along the coast a number of covered tanks will collect the treated sewage which will be piped into the hold of the ship at predetermined periods, taken out to sea, and dumped where currents provide dispersion.

A reliable system of destructive air deodorization is required on land as it is imperative to control odour emission from the storage tanks as well as from the ship's holds under conditions of charging. The air expelled from the holds by the inflowing sewage is highly offensive. To avoid the necessity of deodorizing this volume of air separately, it will be piped back into the tank automatically by the negative pressure caused in the tank during the period of discharging into the ship. The air from both ship and tank will be deodorized by the equipment installed at the top of the tank, and released to atmosphere.

It is true to say in many cases that air polluted over a sewage works is a greater nuisance than the liquid waste discharged into the sewers. In fact, the final effluent from a modern sewage plant built and operated on scientific lines is clear, free from taste, odour, and microbes, cannot be distinguished from ordinary drinking water—and is as potable. Owing to the increasing use of water for industrial purposes in the face of limited, or shrinking supplies of it, this grade of sewage effluent can well be returned for industrial usage (183). The re-use of purified effluent has been discussed by STANBRIDGE (179) at the Blackpool Conference of the Institute of Sewage Purification. The amount of water involved in this process is illustrated by the capacity of the largest and most up-to-date sewage treatment plant in Britain, the Southern Outfall Works, London, which treats 216 million gallons (nearly 1000 million litre) of sewage per day (180). Other sources of industrial water are desalinated sea water (181) and process water (182).

This is an indication of the trend to save raw materials by converting waste into a useful raw material. Destruction or dumping of waste is the end point of a straight line process whereas conversion of waste into new material makes the flow line circular. This is a highly important aspect economically and ecologically as it helps preserve natural resources which never are unlimited how much man may please himself in thinking otherwise.

Synthetic detergents have been, and sometimes still are, a menace to the proper operation of a sewage treatment plant. The days are still vivid in every sewage works manager's mind when detergent bubbles piled high on tanks and lagoons, and froth was blown by every breeze all over the place, including neighbouring fields and pastures. The first detergents appeared to be almost indestructible, chemically. Later, their biodegradability was improved by using, or developing, new base materials. Of these, linear alkylbenzene sulphonates (straight chain alkanes, n-paraffins), stand in the forefront, and of the non-ionic detergents those based on alcohol ethoxylates, which are biodegradable, are of prime importance. The linear alkylbenzenes have side chains containing molecules each comprising 10–15 carbon atoms. The bacterial degradation of straight chain alkanes enables the production of synthetic protein foodstuffs from petroleum fractions.

The little or non-foaming detergent is greatly preferable in sewage treatment. Non-ionic surface-active agents are promising, especially as their foaming power can be controlled by suitably modifying their ethylene oxide content.

Detergents have often disrupted the proper working of sewage treatment works by inhibiting sludge digestion. It is therefore essential to assess the biodegradability of synthetic detergents, and a proper test method has been worked out at the Water Pollution Research Laboratory (184). The effluent problem of detergents is specified in terms of biochemical oxygen demand. A test for the BOD of detergents in connection with sewage and effluent problems is described in BS 3762.

An investigation of soil pollution by faecal matter (187) showed that within 'smelling distance' soil pollution was high. Air pollution was considerable within some 600 m (2000 ft) from a pig-breeding farm, but only 200 m (660 ft) around a dairy farm. These distances have been suggested as minimum safe distances between agricultural land and residential districts to prevent polluted soil in the latter area.

27 Pharmaceutical industry

In this industry the ideal factory layout is approached. The plant is kept scrupulously clean, and all processes are closely controlled. Processing vessels are of the closed type, and materials flow in pipe systems. The products, byproducts, and waste effluents are osmogenic in varying degrees, and problems arise from limitations imposed by the process or material. In such instances, alterations of technique, perhaps only slight, will accommodate all restrictions, and produce the desired effect of odourlessness, or air deodorization.

An instance was afforded by the fermenter section in the production of antibiotics. A pilot plant ran small batches of the same characteristics as the production material, and it was reasonable to assume that any findings on deodorization would be applicable to the effluent vapours from the production fermenter.

On inspection, and carrying out several smell tests with the effluent vapours released at roof level, both odour and taste were found to be of the protein type, rather like slightly overdone porridge; a fine whitish-grey precipitate lined the duct from the cyclone to the roof; and the condensed steam waste contained a suspension of fine particulate matter not below the limit of visibility in normal daylight. The odour seemed to have a certain persistency, to be of high molecular weight, and heavier than air. This was confirmed by people living nearby, although the vent shaft was some 40 ft (13 m) above ground. Neither high buildings nor lines of tall trees intercept the travel of the polluted air. The particulate matter in the steam condensate was solids from the fermenter, entrained in the air, and fouling the sewerage system.

Deodorization of air was based on a choice of two methods: storing the smell by adsorbing it on beds of activated carbon; the other, destroying the smell by irradiation with ultra-violet rays. Tests showed the superiority of the latter method and this was adopted for full-scale operation.

28 Oil refineries

The petroleum industry encompasses such a wide range of activities that it is impossible to give here more than a glimpse of the many lines of production. Common to all of them is operation in completely closed systems, an ideal condition for securing an atmosphere free from pollution. Environmental air will become polluted only if a closed system springs a leak, or worse. Many processes have been revised and newly formulated resulting in greatly reduced, or altogether absent, waste or byproducts.

The petroleum industry is not only concerned with the refining of crude oil from which to produce the various grades of fuel oils, but also with manufacturing the petroleum chemicals which represent one of the largest capital investments in the chemical industry. Characteristic processes are the production of: ethanol C_2H_5OH, from ethylene C_2H_4; styrene by alkylation of benzene C_6H_6 with ethylene and dehydrating the intermediate ethylbenzene; alkenes by cracking liquid hydrocarbons. Amongst a long list of chemicals, all derived from petroleum, are members of the methane series, of the ethylene series, acetylene, aromatic hydrocarbons, and synthetic plastics and rubbers. Many of the products are of a highly offensive nature, such as the oxidation of propylene to form acrolein, from which is derived allyl alcohol, and ultimately, glycerol; but it also serves in the synthesis of methionine, $CH_3S.CH_2.CH_2.CHNH_2COOH$, an amino acid, by interacting acrolein with methyl mercaptan.

Oil refineries, being a chemical plant, must in the UK conform to the stipulations of the Alkali Act, 1906, and the Alkali Works Order, 1966. Suggested limits of emission and a basis for determining heights of gaseous effluent discharge were discussed at the International Clean Air Congress, London, 1966 (231). In all chemical works producing more than one commodity there is danger of the gaseous effluents from the various processes interacting in free air, through carrying liquid or solid particulate matter from the processed material. It is therefore a primary consideration in siting an oil refinery (emitting acid gases), to keep it away from works producing ammonia or products containing ammonia, e.g. fertilizer, and so to site both plants that the prevailing wind will not—so far as can be foreseen or calculated—mix their gaseous emissions. If they should get mixed, a white stable cloud will result from chemical interaction, and may drift for miles.

Osmogenic emissions from oil refinery stacks will cause several types of smells: oily odours; acrid odours from flue gases; and sulphurous or sour odours from sulphydryls and hydrogen sulphide. Acrid fumes are released at chimney mouth level and the winds carry them rapidly away, spreading and diluting the offensive waste down to acceptable limits. Oily and sulphur smells are released near the ground. They are at ambient temperature, hug the ground profile, and may roll along over miles of countryside because they are mostly heavier than air. They are also the more noticeable because they stay on the ground where there may be little air movement. Conditions which prevent air movement or turbulence near the ground will also prevent the dispersal of the cold (oily and sour) emission in contrast to the hot (acrid) emission from the chimney tops. Stable warm days and temperature inversions will keep the cold cloud stationary for the duration of the meteorological condition.

Sources of offensive smells in refineries are well defined and therefore easily identified in cases of mishap. They are: storage tanks, polluted waste water, gas emitted from the piped system (flares), and accidental leaks of liquids or gas.

Evaporation from storage tanks may be considerable in hot climates or during hot weather elsewhere. Reduction of evaporation has been partly successful by using infra-red reflecting (aluminium) paint on the outer walls and the tops of tanks. A much better solution is the floating roof tank where the roof is directly supported by the free surface of the contents which now are 'free' no longer. The roof will prevent evaporation to a very considerable extent and the only molecules which gather sufficient thermal energy to move off independently of their neighbours are those forming a thin film on the inner wall of the cylindrical tank after oil has been drawn off, and the roof has moved downwards. The floating roof tank also has the great advantage that there never is an empty space between oil level and roof as there is in fixed roof tanks, and during filling periods no oil fumes are ejected. On the contrary, filling a floating roof tank eliminates any odour there was from the oil film on the inner tank wall above roof level as it now moves to the top, covering the wall entirely. If fixed roof tanks are used, gaseous effluents and polluted air should be collected from all tanks and conducted to an air deodorization plant.

The sulphuric or sour products are refined by washing and neutralizing with caustic soda solution (sodium hydroxide, NaOH, containing 60–75% of sodium oxide, Na_2O).

If sour water (direct from the washing process, and without having been neutralized by the alkali), or spent caustic soda solution are discharged into the sewers without having been treated to required standards first, most disagreeable smells will be caused and spread throughout the sewerage system. It is usual to reduce these effluents by steam stripping, burning of the gaseous fraction in furnaces, and discharging of the liquids into the sewers.

Emissions of combustible gases are circuited into the flare system, and burnt (317). This oxidizes sulphides (hydrogen sulphide) and sulphydryls (mercaptans). If the temperature of the flare suffices, for instance by adding high grade fuel to the stream, other odorivectors, too, will undergo deodorization by thermal oxidation (combustion). But there are several compounds which will not at all, or only partially, undergo thermal oxidation. The consequent intermediate products released through the flare cause smell; this can be relieved by adding nascent oxygen as a final step after combustion, before the waste gases are released to atmosphere. A survey of techniques available for alleviating the pollution of air and process water by operating olefin plants is given by MENCHER (232).

High sulphur-containing crude oil and natural gas are sources of sulphur which, on recovery from the raw materials, commands a high price. The process described by CHUTE (234) is based on the burning of one third of the hydrogen sulphide to sulphur dioxide. This is then reacted with the remainder of hydrogen sulphide to yield elemental sulphur and water— basically the well-known CLAUS process of partial oxidation. The sulphur is extracted in the form of H_2S by absorbing it in an aqueous solution in ethanolamine. Condensers remove the sulphur and, after passing through a coalescer, the droplets of sulphur conglomerate. The process reduces hydrogen sulphide to well below 10 ppm and releases the waste gases from a high stack for quick dilution so that no odours pollute the air. Sulphur recovery is better than 95%. All water is available in the form of steam needed in the treatment plant.

The height of the stack is all-important. This is proved by using very high chimneys spaced out well. The Ingolstadt (Bavarian BP) refineries have built their works in the midst of tourist country, yet any smells there might be are released at 150 m (500 ft) above ground, and diluted quickly. The works operate on Libyan oil (0·5% sulphur) and Iranian oil (1·2% sulphur) (253). The sulphur-containing gases are collected from the entire plant (hydrogenation, hydrofiner, catalytic cracking) and are washed, then heated. The sulphur contained in crude oil is eliminated by passing the oil through hydrotreaters where hydrogen sulphide is formed. The H_2S is separated, and burnt with insufficient oxygen producing a variety of sulphides. A catalyst transforms the gases into liquid sulphur which is kept heated in storage tanks. Waste gases from this process have no smell and are released to atmosphere. The refinery produces annually some 10 000 t of purest sulphur.

The economics of air pollution from the refineries' point of view have been discussed at the 7th World Petroleum Congress (236). The socio-economic impact of air pollution from petrol exhaust is put into focus by statements such as that the total of extra household cleaning costs due to polluted air in Washington alone is nearly 250 million dollars a year (237). It is understandable that the industry searches frantically for methods which will reduce car exhaust (285). Amongst the many tried more or less successfully there is one recently suggested which reduces hydrocarbon and carbon monoxide content by adding fresh air at the exhaust valve. Unfortunately, it also increases odour intensity quite considerably (238), and is not likely to be used widely for this reason alone.

Methods for the measurement of air pollution around oil refineries were reviewed by the Working Group on Stack Height and Atmospheric Dispersion of the International Study Group for the Conservation of Clean Air and Water (Western Europe) (254). The report offers methods for the measurement of airborne hydrogen sulphide, sulphur dioxide, sulphydryls, nitrogen oxides, oxidants, hydrocarbons, and of smoke, grit, and dust. The permissible amount of dustfall is 150 $t/km^2/year$ (286), i.e. about 0·5 $g/m^2/day$. Gaseous effluents from refinery chimneys should be analyzed regularly by standard procedure.

29 Paper mills

A particularly osmogenic process is the one producing kraft paper, an unsized strong paper made from sulphate paper pulp. This is obtained by digesting wood pulp in a solution of sodium sulphate with some sodium carbonate and sodium hydroxide.

The 'black liquor' left from the process passes through recovery furnaces, evaporators, digesters, lime kilns, oxidation towers, and dissolving tanks, which cause smells in about that order of intensity. The pollutants are hydrogen sulphide, methyl mercaptan, dimethylsulphide, dimethyldisulphide, and sulphur dioxide; amounts of these gases formed at the various stages have been determined (199). Pyrolysis of the liquor showed the presence of well over 60 compounds, half of which were in the gaseous state and of these, again about half, were sulphur-containing derivatives. The pyrolysis products contain some 70 per cent of the total sulphur in black liquor. Quality and quantity of the waste product depends on the parameters of pyrolysis and combustion. Methods of investigation and instrumentation have been described in detail (200).

An examination of exhaust gases and crude terpentine from kraft mills yielded the following result.

Raw turpentine contains: methyl mercaptan CH_3SH, dimethylsulphide $(CH_3)_2S$, propanone CH_3COCH_3 (acetone), ethanol C_2H_5OH, butanone (ethyl methyl ketone) $C_2H_5COCH_3$, trimethylacetaldehyde $(CH_3)_2CHCOCH_3$, 2-pentanone (propyl methyl ketone) $C_3H_7COCH_3$, dimethyldisulphide CH_3SSCH_3, and traces of α-pinene, β-pinene, camphene, and 3-carene, all containing 10 C and 16 H atoms.

The raw turpentine was fractionally distilled, yielding three fractions.

First fraction: up to 50°C.   Methanol, ethanol, water, methyl mercaptan, propanone, dimethyl sulphide;

156

Second fraction: 50–120°C.   Dimethylsulphide, methyl mercaptan,
propanone, 2-methylfuran, isopropanol,
trimethylacetaldehyde, 2-pentanone,
α-pinene, butanone, butanol, 2-butanol,
propanol, dimethyldisulphide, 3-carene;
Third fraction: above 120°C.   α-pinene, β-pinene, camphene, 3-carene,
α-terpinene, γ-terpinene, limonene,
β-phellandrine, terpinolene, p-cymene.

There are several uncondensable gases: methane, ethene (ethylene), ethane, propene (propylene), propane, methanol, methyl mercaptan, dimethylsulphide (201). Analysis was by gas chromatography and, in some cases, mass spectrometry.

From this analysis, which has been reported in detail to demonstrate the complexity of air pollution problems by smell, it becomes clear that there are no easy anti-pollution measures. But British Patent 818 572 (1959) describes relatively simple measures. The sulphurous gases generated during the preparation of paper pulp by the sulphate process are chemically washed in a solution of caustic soda, NaOH, 175 mg/l, in the sump of a wet-scrubber, whence it is pumped back to the top only to make it fall through a perforated plate to the bottom of the scrubber. The liquid is recirculated several times and absorbs the gases completely. It is then used for the production of paper pulp at a temperature of 30–40°C. The repeated recirculation is necessary to allow for the required reaction time.

The technique of deodorization otherwise requires that all processes must be carried out in closed systems so that no vapours or gases can escape. Any process suitable to dealing with sulphides is useful as a means of air deodorization.

30 Other industries

The enamelling and lacquering industry provides a wide range of smells according to the type of solvent used for the various natural and synthetic materials employed in the finishing trade. Solvent vapours are highly volatile, as a rule, and give rise to the formation of photo-chemical smog, i.e., smog constituted of chemicals produced by solar irradiation, under suitable meteorological conditions, of airborne vapours. From work by MAY (160) who investigated 37 industrial solvents Table 27 has been adapted.

It would seem that the uncontrolled emission of solvent vapours into the air calls for control by law or by Order of the Local Council. At a Conference in San Francisco, October, 1967, this point was raised and hotly debated. The rapporteur (161) suggested that no common action should be taken on the grounds that the motor vehicle exhaust presents a much bigger problem of air pollution than solvent vapours. To control this relatively minor nuisance could well turn out to be an expense producing little benefit for the public. In crass contrast to this statement is the opinion of GOODELL (273) to the effect that one industrial enamelling oven will produce as much air pollution as some 7000 cars using a total of 560 000 gallons of petrol of which 6–7% leave the exhaust unburnt. A similar attitude is also taken by the Public Health Services of the USA.

The Battelle Memorial Institute has prepared a report (216) concerning the disposal problems of plastics. Although this material represents only 1·5% of the solid waste, but may double before 1970, and although plastics are considered and classified as a nuisance material, no action against its disposal by incineration is proposed. Plastics are halogen bearing, and most of them contain flame inhibitors. Combustion releases hydrogen halides, especially hydrogen chloride which leaves the incinerator in the form of hydrochloric acid. Its concentration may reach the limit of toxicity (maximum acceptable concentration 5 ppm) under un-

Table 27 *Industrial solvent odours (concentrations in ppm)*

Solvent	Maximum acceptable	Distinct odour	Smallest measurable
Light petrol	500	1000	—
Heavy petrol	500	40	—
Heptane	500	320	1
Octane	500	250	1
Benzene	25	90	1
Toluol	200	70	2
Xylol	200	40	3
Methylenechloride	500	220	7
1,2-Dichloroethane	100	180	1
Carbon tetrachloride	10	250	17
1,1,1-Trichloroethane	200	700	3
Trichloroethylene	100	100	3
Tetrachloroethylene	100	70	2
Methanol	200	8800	25
Ethanol	1000	100	4
n-Propanol	—	60	4
i-Propanol	400	50	2
n-Butanol	100	16	3
i-Butanol	—	50	3
Amyl alcohol	100	20	27
Methyl glycol	25	90	23
Ethyl glycol	200	50	10
Dioxan	100	270	7
Tetrahydrofuran	200	60	2
Methyl acetate	200	300	2
Ethyl acetate	400	70	2
n-Propyl acetate	200	25	3
i-Propyl acetate	—	40	3
n-Butyl acetate	200	11	3
i-Butyl acetate	—	7	2
Methyl formate	100	2750	2
n-Butyl formate	—	20	3
Acetone	1000	680	1
Methylethylketone	200	50	2
Diethylketone	—	14	2
Methylpropylketone	200	13	2
Methyl iso-butyl ketone	100	15	2

favourable conditions, but there are no plans to develop air quality standards or criteria for hydrogen chloride. The only correct and acceptable manner of disposing of plastic waste is by controlled tipping (land filling). A detailed survey of the technical and economic aspects of the disposal of plastic waste with reference to osmogenic air pollution recommends stern measures (257). As a means of odour destruction, direct flame incineration in the effluent gases from curing plastics and rubbers is recommended, especially when they contain high boiling point plasticisers, e.g., dioctyl phthalate, diisooctyl phthalate, and mineral oil. Combustion temperature must be at least 570°C (1050°F). By using a suitable catalyst, combustion temperature can be reduced to 510°C (950°F). However, the process is uneconomical (258).

A rather unusual cause of industrial smell is the moulding shop of a foundry. Sand-forming is often done with a mixture of molasses and sand. During pouring, the mould heats up to such an extent that the molasses, a by-product of sugar, being an organic compound, burns up, the results depending on the temperature reached in the mould: initially the products are pyrocatechol, water, and oxygen:

$$C_6H_{12}O_6 \rightarrow C_6H_4(OH)_2 + 3H_2O + O$$

The presence of nascent oxygen will oxidize the pyrocatechol to quinone, $C_6H_4O_2$, which is odourless. At higher temperatures, carbonization produces a blue smoke, sometimes with acetic acid fumes from an intermediate stage.

Part IV

TECHNIQUES OF
AIR DEODORIZATION

31 Phase separation

Pollutants occur in all three phases: solid particulate matter in air—aerosols; liquid particulate matter in air—fogs, mists; and gaseous (molecular) matter in air.

Attention is drawn to an unfortunate mix-up of definitions involving the term *aerosol*. The physical definition is clear enough, explaining it as a colloidal system consisting of two phases, the external phase being a gas and the internal being either a solid, liquid, or gas. In a colloidal system, there is an internal or dispersed phase which is in a state of extreme sub-division (colloidal state), and an external or dispersion phase which is coherent or continuous. Theoretically, there are therefore nine combinations possible, but only eight of them are realized, practically.

| | | *External phase* | | |
		in solid	in liquid	in gas
Internal phase	Solid:	Solid sol	Suspension	Smoke
	Liquid:	Gel	Emulsion	Fog
	Gaseous:	Solid foam	Foam	—
	Generic term	Solid sol	Hydrosol	Aerosol

Among instances may be quoted:

Solid sol:	alloys		*Emulsion:*	crankcase oil
Suspension:	paint		*Fog:*	spray
Smoke:	dust		*Solid foam:*	pumice
Gel:	glue		*Foam:*	lather

The term aerosol is properly reserved for the generic definition explaining that the external phase is a gas. It does not apply solely to the dispersion of a liquid in gas which is, properly, a fog. Commercially, this word is unattractive, hence the general term aerosol is used instead to define a specific system, i.e., a fog.

Before an odour-polluted air stream undergoes antipollution treatment it is essential to remove suspended solids and liquids so that only a gas-gas (purely molecular) mixture remains. The difficulties of separating the various phases grow, on the whole, in inverse proportion to the size of the particulates. There are three classes of treatment at the disposal of the engineer, none of them giving absolute satisfaction. The method most used with solid particles is known as dry-scrubbing or cycloning. Wet-scrubbing is used for the removal of solid and liquid particulates. Foam-scrubbing, a recently-introduced treatment, can be used for either. The separation of a gas mixture into its components or into the carrier gas (here air) and the sum of admixtures (here the osmogenic pollutants) is a difficult matter and often only possible with the use of sorptive materials, such as activated carbon.

32 Wet-scrubbing

The spray nozzle, though still in use today in many a small- to medium-sized plant requiring air deodorization is only of moderate efficiency. The time of contact between the fine water droplets from the nozzle and the airborne solids moving counter-current fashion in the scrubbing chamber, is small and so is the probability that a solid particle will be hit by a water droplet. In counter-current movement the *relative* velocity of the one particle, as seen from the other, is the vector of both velocities, which is quite inadequate. Moreover, when the jet of polluted air rises against the jet of droplets from the nozzle, it curtails the fanning out of the fine spray from the latter, Figure 18a. It is better to introduce a concurrent movement with the water spray at a small angle to the stream of polluted air, Figure 18b. Air velocity should be about the same as the velocity of the water jet at the nozzle so that both droplet of water and solid pollutant sink to the bottom at slow speed which results in long contact time.

Nozzles for the wet-scrubbing of gas should be of the spiral bottom, or of the tangential inlet type. Both have spin chambers and are pressure nozzles producing a hollow cone of fine droplets. Tangential inlet nozzles have orifices up to about 6–7 mm diameter, whereas spiral bottom or ramp nozzles have much larger orifices, up to 40 mm. Droplets are formed within microseconds. The shape of the spray, i.e., the distribution of water droplets, and their range of sizes, determine the efficiency of the nozzle as a scrubbing tool. Below 0.3–0.4 atm (about 5 lb/in^2) water pressure, droplet formation will not occur. Pressure is inversely related to pipe diameter and nozzle orifice. For a nozzle diameter of 1 mm (0·04 in) a pressure of 2·5 atm (about 32 psi) may be applied. Water pressure and nozzle capacity have a log-log relationship.

Nozzles may be arranged to produce sheet patterns of spray (linear coverage), or provide bulk spray (area coverage). In the first case, the nozzles are placed at the apices of equilateral triangles forming a con-

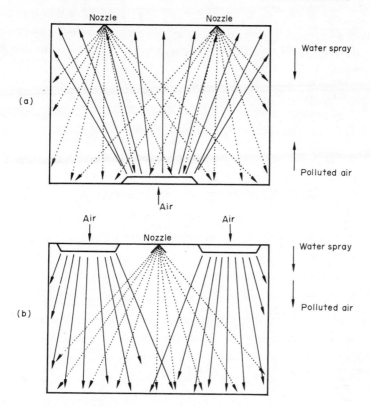

Figure 18   Wet-scrubbing.
(a) Polluted air and water spray move in a countercurrent.
(b) Polluted air and water spray moving concurrent-wise.

tiguous chain. For area coverage, the nozzles should be placed at the centres of contiguous hexagons or their circumscribed circles.

The washed solid particles will form a slurry with the water and may go directly to the main drain if the quality of the waste is acceptable to the Local Authority, or it must be retained in lagoons or settling (sedimentation) tanks above or below ground for a period of time determined by the physical properties of the slurry. The Local Authority acts for the Sewerage Board or the River Board depending on whether the slurry is to be discharged into the sewers or a river. When into a river, the biochemical oxygen demand (BOD) of the waste is the determining factor; when into the sewers, other chemical and biological factors enter into consideration.

The BOD of a liquid, be it industrial waste or sewage, is judged by the amount of oxygen the polluted water will take up in five days when kept in a bottle at 20°C. The BOD is expressed in ppm. An increase or decrease in incubation time or temperature will increase or decrease the BOD, and it is essential that the 5 days incubation time at 20°C, as recommended by the Royal Commission on Sewage Disposal, be observed.

The recommended standard BOD value of domestic sewage is 0·167 lb (74·315 g) of oxygen per head of population, the BOD of industrial waste being expressed in the same unit. Dairy waste, for instance, has an oxygen demand equivalent to that of sewage from 1000–3000 people.

Table 28   *Average BOD values of industrial wastes*

Waste	lb (kg)/day	Population equivalent
Cotton mills	1000 (450)	6000
Dairies	350 (150)	2000
Paper mills	750 (300)	4500
Tanneries	3500 (1500)	20 000

The biochemical oxygen demand of sewage and waste is in competition with the oxygen requirements of aquatic life. Bottom-feeding fish in a stream require oxygen at the rate of approximately 4 ppm, and fast-moving surface fish up to 10 ppm (118). Anaerobic decomposition and subsequent production of offensive smells occurs in rivers whose oxygen content is down to 2 ppm, or less. Uncontrolled discharge into rivers will not only foul, and then destroy, life by means of its chemical composition, but also by reducing the available oxygen.

In wet-scrubbing the solid particles are expected to be removed by precipitation or flocculation (impact precipitation), and the liquids by dissolving in the water spray. Similarly, gaseous matter may be removed by dissolving in the water spray. The latter is difficult of achievement since only a proportion of the constituents of gaseous waste is water-soluble.

Table 29 *Aqueous solubility of some osmogenic substances*

Acetone	∞
Acetyl methyl carbinol	slight
Allyl mercaptan	insol.
Amyl acetate	1·75%
Butyric acid	5·6%
L-cysteine	0·01%
Dimethyl amine	slight
Ethyl acrylate	insol.
Ethyl mercaptan	1·5%
Furfural	insol.
Hydrogen sulphide	430%
Indole	ss
Methyl acrylate	insol.
Methyl amyl carbinol	0·35%
Methyl amyl ketone	0·43%
Methyl ethyl ketone	insol.
Methyl isobutyl ketone	1·98%
Methyl mercaptan	insol.
Phenol	6·7%
Skatole	0·05%
Trimethylamine	insol.
Vanillin	slight

The temperature of the water used in wet-scrubbing need not be raised above normal, i.e., about 5°C. Solubility is inversely proportional to the temperature of the solvent (water).

Table 30 *Solubility and temperature*

°C	H_2S	SO_2	atmospheric	
			N_2	O_2
0	4·67	79·79	28·90	69·82
5	3·98	67·48	25·64	61·21
10	3·40	56·65	22·87	54·30
25	2·23	32·79	17·68	40·73
50	1·39	9·0	13·55	30·20
70	1·02	—	12·49	26·77
100	0·81	—	12·26	25·21

A droplet falling from a nozzle takes on a pear shape. Surface tension makes the interface, i.e., the boundary layer of molecules in the droplet, offer a certain resistance to the ingress of gaseous molecules. If the resistance between the liquid and the gaseous phase is high, the gas will hardly dissolve in the water droplets of the mist. Absorption is a function of interface resistance. An improvement can be achieved by reversing the relationship of the two phases. When the nozzle type of scrubber is used, water droplets are sprayed through the gas in the chamber. Reversing the process means to bubble the gas through a tank containing water.

If required, chemical additives may be selected to neutralize either acid or alkaline wastes so that the discharge to the sewers consists of water in which a salt is dissolved. By various technical expedients the contact time between an individual gas bubble and the scrubbing water can be controlled. The scrubbing tank can be open to atmosphere or be covered and have a safety valve type control element at its highest point. This arrangement will satisfactorily prolong contact time between the two phases, and allow blow-off when a predetermined pressure in the space overlying the water surface has been reached.

Owing to the low odour thresholds, the portion of pollutant in the stream is very small and any equipment designed for deodorization of the air must handle a large volume of carrier gas (air) in order to deal with the pollutant. It should also be realized that any water-insoluble compounds will leave the water spray or the scrubbing tank unchanged except, perhaps, for a proportion which may have been oxidized by passing through the water. The solubles will remain in the water and impart their odour to it. A frequent change of water is essential. This may be expensive and difficult. Since it is still good practice to license an offensive trade only if its place of work is well out of the way, this very condition makes for difficulties in both water supply and drainage.

Under these conditions a recirculation system is often thought of as economical financially and saving water supplies. A recirculation system will generally include a condenser if hot gases or steam are handled, and cooling towers are also provided. These soon become a source of odour if the osmogenic material cannot be completely removed from the water before it reaches the towers. The operating cycle has four stages:

(1) Polluted air is scrubbed: odour is transferred from air to water.

(2) Water becomes tainted, and must be renewed.

(3) Water is deodorized.

(4) Water passes through cooling towers which emit the same smell as originally contained in the air.

Cooling towers are constructed from wood which has been impregnated and, when new, is water repellent. After a few years of operation the wooden structure has absorbed sufficient pollution to become itself a source of odour. Algae often invade the cooling towers, usually green algae (*Chlorophyceae*), i.e., *Chlorella,* a unicellular micro-organism, *Ulothrix,* a filamentous, and *Scenedesmus,* a unicellular alga.

The most frequently found blue-green alga (*Myxophyceae*) is the species *Oscillatoria,* a filamentous organism. Diatoms (*Bacillariophyceae*) also thrive in wooden cooling towers. A number of algicides are available (119), chromate being rather ineffective unless in high concentrations. However, *Ulothrix* and *Oscillatoria* are, if not chromate resisting, fairly tolerant to it, up to concentrations of 120 ppm.

At elevated temperatures the energy of molecules increases, so the retentive power of water droplets diminishes with rising temperature. Diffusivity, the inverse of retentivity, is defined by the Stokes-Einstein law as

$$D = kT(d + 1 \cdot 72\lambda_m)/3d^2$$

where D = Diffusivity, k = Boltzmann's constant, T = absolute temperature, °K, λ_m = mean free path of the gas molecule, and d = diameter of the particulate matter.

When there is no particulate matter (solid or liquid) polluting the air, but an osmogenic gas, d is the diameter of a gas molecule, and diffusivity becomes a maximum since a molecular diameter may be assumed of magnitude $d = 3$ Å $= 3 \times 10^{-8}$ cm which is much smaller than even the smallest colloidal particle ($10 - 1000$ Å $= 10^{-7} - 10^{-5}$ cm). Diffusivity is inversely proportional to d^2.

Structural materials for cooling towers other than wood are aluminium, 18-12 molybdenum steel, titanium, Hastelloy C (high nickel content alloyed with iron, molybdenum, silicon, aluminium), Monel, and Inconel (nickel alloys) (120). Concrete has also been used a great deal. Durability depends on the anticorrosion properties of the structural materials, and the application. Metal parts, or even the entire tower, may be coated for protection with Neoprene (a brand of polychloroprene, PCP). This can be used at temperatures up to 120°C, and is chemically resistant to oil, grease, weathering, direct sunlight, ozone, and to aliphatic hydrocarbons.

The process of oxidizing any suitable pollutant by bubbling it through a scrubbing tank is aided, according to the Norwegian Patent 92 995 (ref. 121), by acidifying the water and electrolysing it. In coastal areas sea-water is used for the purpose, but inland, ordinary salt or hydrochloric acid may be added instead. Since the process of destructive deodorization depends on the presence of nascent oxygen (from the water) or nascent chlorine (from the hydrochloric acid or its salts) it is obvious that sulphuric acid would be useless as an acidifying agent. It should be remembered that nascent chlorine does not really deodorize as a primary agent, but that its action is represented by

$$H_2O + Cl_2 \rightarrow HCl + HOCl \rightarrow 2HCl + O$$

The use of sea-water as an industrial coolant, and the selection of the proper structural materials, is discussed by BROOKS (314). Sea-water consists mainly of chlorine, sodium and magnesium, the latter being responsible for the bitter taste.

Table 31   *Analysis of sea-water* (122)

Chlorine	18 980 mg/kg
Sodium	10 561
Magnesium	1 277
Sulphur	884
Calcium	400
Potassium	380
Bromine	65
Carbon	28
Strontium	13
Boron	4·6
Fluorine	4·6
Silicium	4
Some 28 elements below	1

The Norwegian patent is particularly applicable to the treatment of sewage turned septic. Anaerobic bacteria are also destroyed thus inhibiting the production of hydrogen sulphide. This gas is the result of biochemical reactions involving the life process of micro-organisms belonging to the genera *Spirillum* and *Microspira*. They are the species *Vibrio desulfuricans, V. thermodesulfuricans,* and *V. estuarii* which act on sulphates, for instance, $Na_2S_2O_3 + H_2 \rightarrow Na_2SO_3 + H_2S$.

The application of bacterial activity to the purification of water contaminated by sewage is well known to the specialist. A similar process is possible with reference to polluted air. The system has been developed by Babcock & Wilcox and consists of huge trays on which a suitable carrier material is placed. This is inoculated with the bacteria most appropriate to digesting the osmogenic gas. The filter beds are so arranged that no air can bypass them, and sufficient surface area and thickness of the deodorizing medium must be provided to give a satisfactory contact time with the air passing through the medium at controlled velocity.

Scrubbing towers are different from, and yet similar in general operation to, cooling towers. In scrubbing towers, the scrubbing water, with or without chemical additives, such as caustic soda for the absorption of gaseous phenols, is lifted to the top and descends to the bottom over a series of trays which break down the flow of continuous water filaments into a multitude of individual droplets of various sizes. The trays are arranged in banks so that each drop, falling on a tray below, coalesces with others. Yet, when it reaches the edge of the tray it is broken down again in fine droplets which fall on to the next tray below, and so forth. The waste gas is admitted from the bottom of the scrubbing tower and rises counter-current fashion to the top where it should arrive clean of any particulate or molecular matter it has carried, and escape to atmosphere. The repeated breaking-up and reforming of water droplets is essential because, in the process, new molecules will appear in the interface and exert an attractive force uninhibited by molecules already adhering, or absorbed. The life-time of a droplet descending, in ever changing form, from the topmost tray to the bottom is much longer than in a chamber fitted with spray nozzles, and the overall effect is much improved.

If sorptive power and contact time should still be insufficient, the trays may be substituted by shallow boxes containing special material in a form which causes the water to flow a longer way before passing on to the next box below, and sometimes of a kind with improved retentivity. Such material is available in shapes ranging from simple rings to complex saddles and is known under the names of the designers (Raschig rings, Lessing rings, Ross partition rings, Berl saddles) or by the structural characteristics (spiral rings, Intalox saddles). The material may be ceramics, plastics, or metal. Others are small-mesh wire gauze rolled into cylindrical shapes (Dixon gauze packing), or the Panapak or Spraypak contact materials.

Packing materials made of plastics can be shaped to produce a multiplicity of interlocking contacts which are responsible for considerable interstitial hold-up time independent of flow conditions. Standard polyethylene, usable up to 65°C, and high-density polyethylene, up to 110°C, are physically suitable and economical materials. Rose and Young (123), and Leva (124) give detailed discussions of the hydraulic characteristics of packed towers operating under counter-current flow conditions.

The packing material described so far provides, mainly, for an increase in contact time. If, on the other hand, adsorption of the airborne pollutants is required, neither metal nor plastics, not even ceramics are fully satisfactory and a highly porous material, offering a very large active surface in a relatively small volume, should be selected. Activated carbon (see later) suggests itself readily for the purpose but a new slant has been put on the techniques of applying it by immersing the carbon in the liquid (125).

In a cheaper and much less effective version coke breeze has been used for many years instead of carbon by small plant operators. While coke is capable of removing quite useful quantities of odoriferous pollutants from the air, a problem arises the moment the coke has been saturated. The adsorbed matter can be removed from the coke breeze by washing it with chlorinated water, but this indicates immediately the shortcomings of the method. A full scale chlorination plant must be kept at the ready to operate at irregular intervals. The airborne smell is first transferred to the coke, then to the scrubbing water, and is then destroyed by chlorination, i.e., oxidation. Coke breeze may nevertheless prove its worth in emergencies, but then chlorination may not be available. In this case recharging the scrubber with fresh coke breeze and disposing of the saturated material in a sensible way—not by burning it—is the recommended process. Burning the coke saturated with osmogenic matter is a bad policy, since the heated coke will release the odorivectors long before it reaches combustion temperature.

There are other methods for the wet scrubbing of gases, in which no packing materials are used, but there is a zone where the liquid and the gaseous phases meet under highly turbulent conditions. This makes for intimate and long contact between the phases. In some scrubbers the gas moves in a downward direction while the scrubbing liquor moves towards the top. The *cyclonic spray tower* contains an array of nozzles rotating at high speed in a cylindrical tank (Pease-Anthony scrubber). Finest droplets are produced forming an effective mist in the tank. The mist moves centrifugally from the axis of the tank, and the air is admitted tangentially meeting the spray from the rotating nozzles head-on. It is also possible to keep the nozzles stationary and to move the air at speed. In either case, it is essential to produce a strong impact between the phases. Large volumes of air can be handled by a cyclonic spray scrubber, exceeding even 60 000 m^3/min at 10 atm pressure

A *fog-filter scrubber* produces tangentially directed water sprays from fixed nozzles in a tank. Air is admitted in a tangential direction, too, yet the air is blown into the face of the water jets. Very large volumes of air at pressures up to 40 atm are handled by fog-filter scrubbers. The principle of these scrubbers is the sudden arresting of movement of a particle by causing it to collide with another travelling at speed in the opposite direction, for instance as described in British Patent No 1 021 573 (1966).

In wet-scrubbers impact is achieved between a liquid phase (water droplets) and a solid phase (osmogenic particulates). The principle has also been applied to *impact scrubbers* where polluted air enters at speed a chamber filled with solid obstacles, for instance, strike plates. These plates are so arranged that the air impinges upon them with force, thereby depositing any airborne matter on the plates. A thin layer of water is made to run down the surface of the strike plates cleaning them continuously and sweeping the deposited matter into the sump. To increase the effect, the baffles are deeply corrugated, which increases their surface. The air strikes the baffles at an angle. The sheet of water remains coherent owing to surface tension and impact occurs between the air and the film of water moving over the plates. The corrugated baffles may be of any suitable material (stainless steel, aluminium, asbestos cement) and act merely as structural support of the film of water.

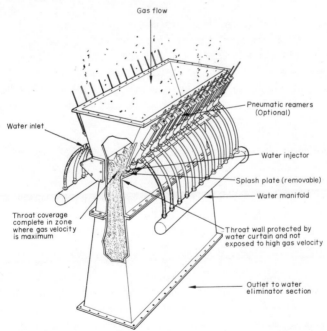

Figure 19   The Kinpactor, a wet-scrubber using the venturi principle. (*Courtesy American Air Filter Co.*)

Another design of impact scrubber is shown in Figure 19. The function is based on the principle of the venturi tube, i.e., the conversion of pressure energy into kinetic energy. The polluted gas stream is pressurized by a suitable fan and directed at the throat of the venturi where conversion of pressure into velocity takes place. Sheets of water are also injected into the throat of the venturi, where the polluted air stream atomizes the water. The impact of the high velocity particulates on the low velocity mist results in a thorough precipitation of all solid matter. The fact that pressurized water is used in sheets rather than as a solid jet or atomized spray eliminates the chance of untreated air bypassing the water. The sheet of water must completely fill the throat of the venturi. The throat should be the narrower, the smaller the size of the particulate matter, thus increasing the velocity of the polluted air stream. This is a necessary requirement since with wide throats and low velocities the pollutant particle tends to be pushed by, and around, the water droplet instead of striking and penetrating into it. The variability of throat width and inlet pressures results in satisfactory efficiencies for any size of particle within the design range of the impact scrubber.

An impact collector is shown in Figure 20, having large wetted surface areas for the impingement of particulate pollutants, and a self-induced spray with water supplied through a non-clogging pipe. A design by HOLMES-SCHNEIBLE, Figure 21, for the removal of particulate matter enables the wet-scrubbing of stack gas emissions to be carried out at source, i.e., at the top of the stack. The hot, polluted gases are burnt in the upper portion of the cupola stack so that all carbonaceous matter is converted to gaseous products which are then washed and cooled in the converter and, finally, released to atmosphere.

The collector consists, principally, of an outer shell with collection trough. The stack gases, once in the cupola, are confined to an annular passage between the shell wall and an inner cone over which water flows across the passage. Gas moving through this water barrier will be stripped of all particulate matter and leave the cupola clean and inoffensive. Removal efficiencies are better than 99 % for particles of sizes greater than 4 μm.

Gaseous osmogenic pollutants in the stack gases are removed if they are easily water soluble. Contact time between the water curtain across the annular passage and the stack effluent is, by necessity, short. The impact collector has not been designed to deal with odoriferous matter in the molecular state.

A combination of bubble tank and impact scrubber yields the *turbulent wet-scrubber*, US Patent Nos 2 621 754 and 2 720 280 which, originally,

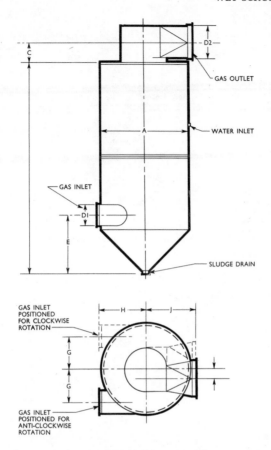

Figure 20   Drawing of a type 'JC' multi-wash collector, and sectional view.
(*Courtesy W. C. Holmes & Co. Ltd.*)

was known as the Doyle scrubber. The great advantage is the high
efficiency and the absence of moving parts. The air inlet tube reaches a
distance into a given volume of water, the distance being proportional to
the pressure of the incoming air, Figure 22. As the air leaves the tube it
will hit the solid block of water, thereby precipitating any particulates
which are kept moving by the turbulent water. Pollutants up to 50%
concentration can be kept suspended in the water. Atomization and

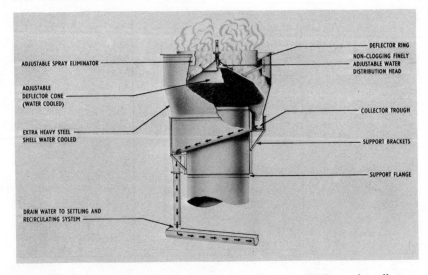

Figure 21   Sectional view of the Holmes-Schneible type 'SW' cupola collector. (*Courtesy W. C. Holmes & Co. Ltd.*)

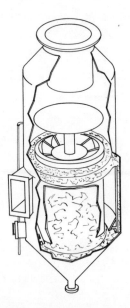

Figure 22   Wet-scrubber without nozzles. Dirty air enters from the top, hits the water and deposits the dust in the water. The clean air leaves the scrubber by the air outlet at the left. (*Courtesy Joy Manufacturing Co., Turbulaire Scrubbers*)

coalescence of the water at the point of air impact is continuous with but little water being lost in the process, mostly by evaporation, and for making-up. Requirements are less than 10 litres of water per 1000 m^3 of polluted air.

Foam equipment for the separation of airborne dust has been developed in Czechoslovakia (127, 128). Two or three perforated trays are arranged in a tank open to atmosphere. The foam is formed in these trays through which the polluted air is fed from the bottom. It is arranged for the air current to enter the tank in a direction towards the bottom sludge receiver, but once in the tank the air flow is reversed and moves through the foam trays. It then passes through a droplet arrestor, and out to atmosphere. The reversal of flow in the tank has a cleaning action in so far as many airborne particles drop out there and into the bottom sludge receiver. The water is let in at the level of the topmost tray, passes through the perforations to the next tray down, and so on if there are more than two trays.

The two factors governing the effectiveness of the arrangement are the turbulence obtaining in the foam layer, and its height. A proportion of the dust is collected below the tray, the rest in the foam. The finer particles are found in suspension below the tray, the coarser ones in the foam. The foam dust separator is most suitable to retain particles of size 2–5 μm and is effective in removing osmogenic dust from rotary kilns and grinding plants. The performance of the equipment is not affected by variations in throughput if changes are kept within 25 % of nominal capacity which is 1600–26 000 m^3/h.

33 Dry-scrubbing

Phase separation by wet-scrubbing removes the pollutant by absorption or precipitation. The scrubbing phase, water, is used either plain or with chemical additives, and requires continuous or periodical replacement and disposal, or recirculation.

Phase separation by dry-scrubbing offers certain advantages. Although the word 'scrubbing' infers the action of 'rubbing hard to clean', especially with soap and water, it has been used a long time by engineers to indicate a cleaning action generally without referring to a liquid medium in particular. It is intended to continue this usage and to define the term as separating pollutants—provided they are solid or liquid particulates—from air without using a liquid phase.

Purification of air from pollutants, whether particulate or molecular is achieved in dry-scrubbing by means of a specially prepared solid phase such as a filter, molecular sieve, electrostatic precipitator, and others. Dry-scrubbing, if applicable to a given process, has the advantage of considerable economy over liquid scrubbing.

Oil impregnated cloth or felt filters are the common type of dry-phase separators. The material may be natural or man-made fibre, mostly glass fibre. The requirements for a filter of this type, generally known as dust filter, are contradictory. Low impedance, i.e., open weave, to keep fan size and power down, small pore size (high impedance) to retain dust particles above the limit of the pore diameter. Compromise results in a thick layer of several thin mats of an open weave which offer but little resistance to the flow of air, yet reflect and deviate the air current many times so that a maximum of fibres is hit. The dust is retained in the oil film covering the fibres. If the pollutant is dust, the Particle Atlas (210) is an essential help in identifying the particulate matter, and tracing it to its source. The Atlas has been described by one of its authors (211). Several other such works are available for the study of special dusts.

Fine wire gauzes of different mesh numbers instead of fibre mats can be arranged to form filters of between 5 and 50 mm thickness.

Glass fibre filters are produced by arranging fibres with diamaters below 10 μm and not less than 0·5 μm to be laid randomly in layers, when heat is applied and the fibres fuse together at their points of contact. A similar process can be applied to metal fibres or metal wires. These filters are useful at high temperatures (glass up to 550°C, metal up to about 200 C deg. below its melting point).

Filters become saturated after some time and should be washed at regular intervals depending on usage. The pregnant solution is led to waste without further treatment.

The filter must be of sufficient mechanical strength not to cave in under the impact of air driven by the fan. Assuming the filter to be rigid and perpendicular to the air flow, the pressure exerted on the filter is

$$p = 2wv^2/2g = wv^2/g$$

where p = air pressure, v = velocity of air, g = gravitational constant, w = weight of dry air.

The weight of dry air may be taken as:

At 0°C (32°F):   1·2936 Kg/m^3 (0·0807 lb/ft^3)
At 17°C (62°F): 1·2819 Kg/m^3 (0·0761 lb/ft^3)

An approximate formula useful to the engineer is

$$p = v^2/430 \text{ (lb-ft-sec system) } (p \text{ in lb/ft}^2; v \text{ in ft/sec})$$

or $$p = 1·1 \times 10^{-5} v^2 \text{ (metric system) } (p \text{ in atm}; v \text{ in m/sec})$$

The size of particulate matter which a fibrous filter will most readily allow to pass is given by

$$a^2 v_m = 10^{-9·4} \text{ cm}^3/\text{sec}$$

where a = the optimum radius for penetration of the particle, expressed in μm and v = mean air velocity in the filter, measured in cm/sec (129).

For highest efficiency of a filter the ratio of active to geometrical surface should be a maximum. This is achieved by folding the filter material in Vee-shape, thus getting about a five-fold increase in surface size. Initial

Table 32   *Pressure exerted on air filters*

Air velocity ft/sec	Air pressure lb/ft²	Air velocity m/sec	Air pressure atm × 10⁻⁵
10	0·23	3	10
15	0·52	4·5	22
20	0·93	6	40
25	1·46	7·5	62
30	2·09	10	110
40	3·74	12·5	172
50	5·81	15	250
75	13·05	20	440
100	23·21	25	690
		30	990

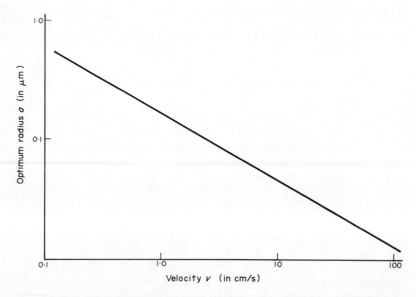

Figure 23   The size of particulate matter which a filter will pass is a function of the optimum radius for penetration of a particle and of the mean air velocity in the fibrous filter.

Table 33 *Pollutant concentration and filter material*

Concentration mg/m^3	Most suitable filter material
5×10^{-1}	Viscous filter
10^{-2}	Cloth
10^{-3}	Fabric
10^{-4}	Asbestos fibre
10^{-5}	Glass or Cellulose fibre or Paper

pressure drop should not exceed 6 mm w.g. (0·25 in w.g.) depending on material and design. A filter is considered saturated when the pressure drop of its single elements (if a multiple filter) has reached three times the initial value.

Penetration of a filter is the most important single factor for assessing its quality. Penetrability is a logarithmic function of the number of particles that have passed through the filter.

Table 34 *Penetrability of filters*

Penetration	Retention	Power of penetration	Log power
100 000	0	10^5	5
10 000	90 000	10^4	4
1000	99 000	10^3	3
100	99 900	10^2	2
10	99 990	10	1
9	99 991	$10^{0·95424}$	0·95 424
8	99 992	$10^{0·90309}$	0·90 309
7	99 993	$10^{0·84509}$	0·84 509
6	99 994	$10^{0·77815}$	0·77 815
5	99 995	$10^{0·69897}$	0·69 897
4	99 996	$10^{0·60206}$	0·60 206
3	99 997	$10^{0·47712}$	0·47 712
2	99 998	$10^{0·30103}$	0·30 103
1	99 999	10^0	0·00 000

The effectiveness of filters is compared by counting the number of particles which have penetrated the barrier. Take, for instance, two filters the one rating at 99 998 retained particles out of 100 000, the other 99 992 particles. The difference of retention is a mere six particles per 100 000. If, however, the number of particles is counted that have penetrated the filter, then it is 2 in the first, and 8 in the latter case. The former filter is four times as effective as the latter.

For testing of efficiency and resistance of filters BS 2831 should be consulted. Penetrability is expressed in terms of a standard dust, i.e., methylene blue for which refer to BS 2577. The size of the dye particles is within the most penetrating range, i.e., $0.2-0.4$ μm. The relationship of these quantities has been critically discussed by HEYWOOD (130).

When a filter has become saturated with airborne particles the engineer is faced with the problem of reconditioning it which means removing, and safely disposing of, the accumulated matter without causing offence anew. Fabric and cloth filters should be washed and may have to be scraped. This is an important procedure if the residue is useful as a waste or valuable as a process material. If, otherwise, complete combustion is attainable at the furnace temperatures, burning of filter and residue is the most economical disposal in certain cases. If only partial oxidation resulted, the furnace would not destroy the smell of the residue, but complicate it.

The removal of hot fat or oil from an air current requires special grease filters which are built on the same principle as dust filters, only they use metallic filter pads of various gauges, coarse and fine, so that the larger droplets are filtered out first. All filters should be placed as close to the source of pollution as possible and certainly before the air is permitted to enter ducting. Filters, but especially grease filters, should be easily exchangeable for cleaning. Grease filters are bathed in hot caustic soda immediately after close-down of operations, at least once daily. A closely woven metal filter also acts like a Davy miner's lamp gauze, preventing a flame from being transmitted across it. New grease traps or filters cause but a small pressure loss of the order of $0.02-0.10$ mm w.g. for air velocities between 100 and 400 ft/min ($0.00\,005-0.00\,025$ atm for air velocities between $0.5-2$ m/sec).

If the hot air is collected by a hood extending over the source of odour, and continuing into a duct it is usual to allow about 30 m^3/min for each square metre of projected hood area (about 100 ft^3/min for each ft^2 of projected hood area). For hoods collecting air heavily polluted with fatty matter in droplet form, such as over grills or fish friers, these figures are increased by at least 50 %. If the grill or frier is given to produce sudden bursts of polluted air the hood should be given sufficient depth so as to accommodate the momentary surplus of hot air before the fan can deal with it. A depth of at least 50–60 cm (2 ft–2 ft 6 in) is indispensable.

Molecular sieves of a structure which allows their industrial application have been developed during the last few years, one of the foremost and first developments being the cross-linked dextran molecular filter which is granular, hydrophilic, and insoluble (131). Further development (132) has produced different types of which the Linde 13X is particularly suited to removing organic gaseous pollutants from an air stream (133). Concentrations of various hydrocarbons from 5–100 ppm flowing at 30 m/min (100 ft/min) are completely removed by the 13X filter. Its denomination is derived from the fact that the cross-linked dextran $(C_6H_{10}O_5)_n$, forming a three dimensional network of polysaccharide chains, encloses hollow spaces (pores) between them which will transmit molecules not greater than 13 Å diameter, but retain all others. The molecular sieve most useful for this application is a synthetic sodium zeolite producing uniformly sized intercrystalline voids amounting to about one-half the total volume. The area-to-mass ratios of 13X are 1–3 m^2/g for the external surface, and about 750 m^2/g for the internal surface of the pores. This compares well with activated carbon although the highest grades of it reach a ratio of 15 000 m^2/g. These are particularly suitable to sorbing molecules below 15 Å size.

Gel structures, such as in molecular sieves, have the disadvantage of collapsing during the process of drying. A newly produced material, stable and permanent under conditions of drying, is an ion exchange resin known as macroreticular resin, ideally suited to serve as a catalyst and adsorber (134).

The action of filters is purely quantitative. A filter will pass all and any particles below a specific size and retain all others. The filter is non-selective regarding other, especially chemical, characteristics of the particulate matter.

Terminology according to MCBAIN (135):

Sorption is the generic term for absorption, adsorption, and persorption
Absorption indicates adherence of the sorbate to the interior of the sorbent
Adsorption is used when the sorbate adheres to the exterior surface of the sorbent
Persorption describes the uptake of a gas on the internal surfaces of a porous sorbent such as activated carbon, silica gel, pumice, and others
Sorbent is the material to whose surfaces the
Sorbate, for instance, gas molecules, adheres.

Selective retention is achieved when persorption replaces filtration. Persorption requires the sorbent to be porous and have a high internal surface-area to mass ratio. This can be increased by activating the sorbate, e.g., by heating to temperatures up to 1000°C. Heating is carried out in various atmospheres, such as steam, air, chlorine, carbon dioxide, any inert gas, or in a vacuum. Maximum activation is obtained by heating in steam; the other atmospheres are listed in decreasing order of effectiveness of activation.

Activated carbon has little physical affinity for hydrogen, nitrogen, and oxygen, in this order, but hydrocarbons are eagerly taken up. Foreign molecules penetrating into the lattice of the carbon atoms are held captive there only if the size of the foreign molecules is below the interatomic distances characteristic of the carbon. In the molecular sieve these 'smaller' molecules or atoms are allowed to pass through the sieve whereas they are retained in activated carbon. In either case, molecules larger than a given threshold size are not admitted to the inner structure of the media. By heating the carbon or charcoal, molecules within the atomic structure are released and driven out thus apparently increasing the surface-mass ratio of the sorbent. Re-activation means the steaming of carbon saturated with foreign molecules or atoms, thereby driving out the foreign molecules. Steam is the safest medium since carbon has but little physical affinity for hydrogen and oxygen, but hot air (oxygen and nitrogen) can also be used with good effect. From Tables 35 and 36 it will be seen that the volume of a gas persorbed on a sorbent differs with the chemical structure of both. For hydrogen, nitrogen, and oxygen, activated carbon has the least sorptive power.

Odour control by sorption includes: adsorption on activated carbon (336); adsorption on other adsorbents; adsorption on impregnated activated carbon (for reactive chemicals in low concentrations which are poorly adsorbed; for instance, ammonia); and chemical reaction with granular reactive material.

Activated carbon adsorption is not in itself a method of disposal, but a means of concentrating the offensive material so that it can be disposed of, or recovered. The design of the system must use the optimum type of activated carbon in a bed thickness and with air velocity giving suitable 'residence time'. This may often be only of the order of 1/50 to 2 sec, which permits the use of compact equipment. The system may provide for in situ regeneration of the charcoal or easy replacement of it. Consideration should be given to potential corrosion, and to economical maintenance.

Most systems based on adsorption use activated carbon, which can take up (and sometimes recover) a wide variety of gases, vapours and liquids. Dust or other particulate matter should be removed by a filter or other device before the air is allowed to pass through the charcoal bed. Although a deep bed of carbon may sometimes give complete and adequate purification, it may be more economical, particularly on a large scale, to do the job in several steps. A cyclone separator will remove coarse particulate matter, a water scrub fine particulates and visible smoke, and a carbon bed organic vapours and gases.

Several types of cell containing activated carbon are used in the control of air pollution, depending on the nature of the problem, including the composition of vapours, their concentration, variation in concentration and flow, other contaminants, etc.

Activated carbon has a sub-microscopic spongy structure consisting of tiny capillary passages, not greatly larger than the size of the molecules that are adsorbed, giving some 200 million sq ft of adsorptive surface for each cubic foot of carbon. (This structure is created in manufacture by burning out part of the carbon substance to form the internal surface.) It adsorbs from 5 to 50 % of its own weight of the larger molecules, depending upon temperature, pressure, concentration, and design of the system. When the available inner surface has become saturated with adsorbed molecules, the original adsorption capacity can in many cases be regenerated by removing the adsorbed molecules with heat, usually in the form of steam.

The principle of adsorption is easily demonstrated with the small apparatus called a Snifter Set, shown in Figure 24. Simply sniffing the air from the set confirms the removal of odours, and relative effectiveness can be judged by measuring the number of bulb squeezes necessary for the first odour to be driven through.

In many cases, a useful system for eliminating fumes or smoke from the air is the combination of a wet-scrubber with an activated carbon unit. The scrubber cools the air, removes any particulate matter and, in some cases, also removes the water-soluble portion of the vapours, thus lessening the load on the activated carbon bed.

Effectiveness varies for different pollutants. Thus butylcellosolve is difficult to adsorb because of a high boiling point; xylol falls in the same category. Freon and propane are difficult to remove because of a low boiling point. Benzene, ethyl alcohol and acetone are easily removed and recovered. Sulphur dioxide and formaldehyde are difficult pollutants for removal or regeneration unless a reacting (catalytic) impregnant is used.

As in other methods of air treatment, the discharge can be treated close to each point of emission (involving, if there are several such points, installing a number of units), or there can be a common duct leading to

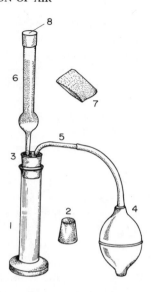

Figure 24   Snifter Set.
1. Glass container to hold odour-producing substance (on cotton wad if liquid).
2. Rubber stopper for container when not in use. 3. Two-hole stopper. 4. Rubber
squeeze bulb. 5. Glass elbow. 6. Purifier tube with charge of activated cocoanut
shell carbon. 7. Replacement charge of carbon. 8. Special single-hole stopper.

a central deodorizing unit. By proper design of the system, the amount of
air required to collect and transport the pollutant is kept to a minimum.

Typical carbon units take the form of steel, stainless steel or plastic
cells containing carbon beds 10–25 mm thick so mounted in frames that
they can be removed and either recharged on site or sent to the factory for
the purpose. Heavy-duty carbon filters of the disposable type may be
returned to the factory for credit, or destroyed on the site by incineration.
The latter method may, however, give rise to the release of incompletely
oxidized (burnt) intermediate products of a smell worse than the original
smell (see Section 38 on Combustion). In most cases it would pay to
return saturated carbon of any description in bulk to the factory for bulk
regeneration. Resorb units and systems are useful where the pollutant
should be recovered because it has some commercial value. The graph in
Figure 25 shows the capital cost of equipment installed to deal with a given
air rate and solvent rate. The operating cost is given for three different sets
of general costs and overheads rated as low, average and high.

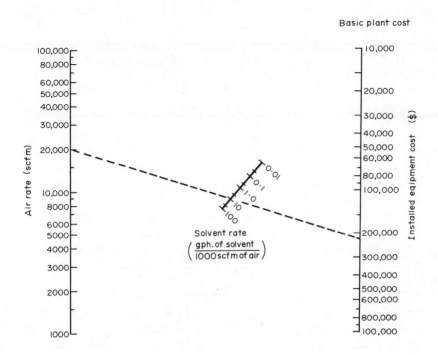

OPERATING COSTS
Basic Costs, Dollars/Gallon

| | Relative cost of operation | | |
	Low	Average	High
Solvent immiscible in water (decantation)	$0·015	$0·025	$0·040
Water-miscible solvent (simple distillation)	0·035	0·060	0·080
Mixed solvent (complex distillation)	0·050	0·080	0·100

Modifying Cost Factors, Dollars/Gallon

	Add	Subtract
Scrubber required	$0·01	
High solvent concentration (0·3 % by vol. or more)		$0·02
Low solvent concentration (less than 0·3 % by vol.)	0·02	

Figure 25   Installed plant cost for solvent recovery.

The depth of the activated carbon layer in a unit depends, in the first place, on the concentration of the pollutant to be adsorbed. For high concentrations, a deep bed, often 500–1200 mm deep, is used, but for low concentrations beds up to 100 mm are more usual. When concentration is low, large volumes of air have to be treated. Keeping the thickness of the carbon bed low, the resistance to air flow remains within acceptable limits and plant cost is reasonable. This applies, for instance, to paint spray booths. By using a thin bed, the cost of the plant and of power can be kept within a practical figure. It is necessary in this case to remove the finely divided paint solids before the air enters the adsorbers, by suitable mechanical filters. The efficiency of removal is ordinarily less in a thin bed than in a thick one, but the lower plant cost is the determining factor.

Rechargeable cells, with large area and flow resistance, are suitable where the pollutants are the ordinary odours of human habitation, e.g. in public buildings, offices or hospitals. They are also used in some process applications and general exhaust systems, particularly where the quantity of pollutants is small, or the discharge intermittent.

A modest-sized cabinet can contain dust filter, carbon filters, and a multispeed blower, for installation, e.g., in conference rooms, laboratories or post-mortem rooms.

Systems incorporating in situ regeneration can deal with higher concentrations. A typical one consists of two or more adsorbers containing activated carbon, connected in parallel to permit single units to be operated alternately on adsorption or regeneration cycles. When the carbon in one unit has become saturated with adsorbed pollutants, the air stream is directed to the other adsorber. The pollutant is removed from the charcoal with steam and the resultant vapour mixture is then condensed. The resulting liquid may be further processed for recovery of the pollutant, or it may be discharged.

If the pollutant is immiscible with water, e.g. hexane, the two layers of condensate are separated in a decanter, the solvent passing to storage, the water flowing to the sewer. In the case of a water-miscible solvent, e.g. ethanol, the condensate passes through intermediate storage to a fractionating column where the water is separated from the solvent. For a complex mixture of solvents, a multi-stage dehydrating and purifying system may be required.

The efficiency and service life of activated carbon adsorber cells can be expressed in a variety of ways, e.g. by reference to the total weight of vapour adsorbed before the carbon reaches saturation. In practice, the service life is difficult to calculate because the fumes consist of complex mixtures. Another approach is to use directly the concept of *odour thresholds* (see Section 3). An instrument for the measurement of odour concentrations by thresholds has been devised and described in the literature (325); it can evaluate odours in both indoor and outdoor situations. Efficiency can be expressed as the percentage of thresholds removed per pass, and service life can be calculated from tables showing the number of threshold cubic units per adsorber cell that can be retained before odour breakthrough, giving such data for types of odour and of space. This method is valuable when the exact chemical composition of the odour is not known; and makes it feasible to develop numerical standards for odour control with either dilution of outside air or adsorption by activated carbon (326).

Several types of continuous adsorption units have been developed and these find applications for specific problems. Continuous systems are, however, not superior to cyclic units for most uses.

Some instances of application may be quoted. In a silk screen printing works, where vapours from solvents used in the inks were lowering efficiency, a system based on activated carbon units so purified the air that it could be recirculated for ventilation. The recovered value of the air (reduced heating cost) together with the improvement in efficiency paid for the installation in a short time. The recovered solvent could be used for cleaning around the plant (327).

A soybean extraction plant uses an activated carbon recovery system for the capture of hexane vapours which were previously discharged directly to atmosphere. The recovered solvent is used in the process and makes the plant safer to operate (330).

Vapours of alcohol and phenolic resins from a factory producing laminated plastics were released directly to atmosphere, losing great quantities of ethyl alcohol each day. Waste air is now scrubbed with an alkaline solution to remove phenolic compounds, and is then passed through two adsorbers in parallel, both containing activated carbon, to remove the alcohol vapours. Periodically, one adsorber is desorbed with steam, while the other, which has just been regenerated, is put back into service. The plant operates the cycles automatically (331).

Special precautions against leaks and high degrees of air purification are required when the vapours to be controlled are not only osmogenic, but also toxic. The pollution must be controlled not at the point of emission, but at the point of usage. Instances include unusual or accidental industrial discharges, severe smog, volcanic action and, in case of war, the use of carbon filters in shelters, factories, and even houses (332).

To determine whether an odour problem can be solved by the use of activated charcoal, the quantity of vapour discharged, and the volume of air required to carry it, are estimated. One of the best ways of measuring the concentration and composition of organic vapour mixtures uses dry ice and acetone in a standard apparatus to freeze the pollutants out of the air stream (333). A method has been published (334) for preparing preliminary estimates of equipment and operating costs for systems for the removal of organic vapours accurately enough for appropriation purposes (335).

Traditional sewage treatment comprises primary and secondary settlement. In the primary step, the heavy suspended matter is settled out, and sludge disposed of, or further purified. In the secondary step some type of oxidation eliminates most of the dissolved organic material. A third step is necessary to make the water suitable for drinking or to meet exacting standards of purification before a Local Authority will accept it into its sewers or river system. This comprises further oxidation, filtration and/or purification by adsorption on activated charcoal.

Solid material, perhaps fermenting or containing components which have sufficient vapour pressure to give off odours, or which are toxic either because of their vapours or because they might dissolve in water and be leached out, can be controlled by the proper use of activated carbon. Such solids, when disposed of by burial or landfilling operations, are covered with powdered or finely granulated activated charcoal which will adsorb gases and fumes, so as to minimize such components reaching the air. Activated carbon will also hold water-soluble components, so that they are either not leached out at all or only at a very slow rate.

The amount of activated carbon for use in such conditions varies in general from about 5 to 50 kg per ton of filling material, depending on the nature and quantity of the osmogenic components.

Table 35   *Sorptive power of activated carbon at NTP*

Sorbate	Volume of sorbate (cm^3) retained per gram of sorbant
Hydrogen	4·7
Nitrogen	8·0
Oxygen	8·2
Nitrous oxide	54
Hydrogen sulphide	99
Ammonia	181
Sulphur dioxide	380

Table 36   *Surface-to-mass ratio of sorbents*

Activated carbon, average	1000 m²/g
Silica gel	614
Chromic oxide gel	185
Porous glass	125
Iron catalysts, average	5

The sorptive power of a sorbent, especially of carbon, can be increased by operating the carbon container at increased pressure and reduced temperature. Low temperature media are:

Nitrogen	−195·8°C
Liquid air	−192
Liquid oxygen	−183
Liquid carbon disulphide	−118
Carbon dioxide (solid snow)	−78·5
Ammonia	−33·35

Sorptive power also increases the easier a sorbate (gas) is transferred into its liquid phase, and the more soluble the gas is in water.

The reason why water is not persorbed on activated carbon, not even by the best grades, for instance, activated cocoanut-shell carbon, is the non-polarity of activated material whereas the water molecule is polar. Water vapour is preferentially sorbed by polar sorbents such as alumina (amorphous aluminium oxide); Fuller's Earth (kaolin containing iron and magnesium); hydrous magnesium silicate; Kieselguhr (diatomaceous earth). These materials are better employed as dehumidifiers and dehydrators than as deodorizers. Non-polar carbon will release moisture previously adsorbed in preference to non-polar (osmogenic) molecules. Activated carbon is a hydrophobic, the others are hydrophilic, sorbents.

Table 37 *Retentivity of activated carbon*

Osmogenic sorbate	Retentivity
Ammonia	L
Body odour	H
Butyric acid	M
Capyric acid	M
Carbon disulphide	M
Cooking odours	H
Essential oils	H
Ethylene	L
Food odours	H
Hydrogen sulphide	L
Indole	M
Mercaptans	M
Putrescine	M
Pyridine	M
Skatole	M
Solvents	M
Sulphur dioxide	L
Tobacco fumes	M
Valeric acid	M
Water (steam)	nil

Note: Retentivity is expressed as the percentage by weight of the osmogenic sorbate carried in dry air at NTP; it is classified as

L	Low	below 10% of its own weight
M	Medium	10–33·4%
H	High	above 33·4%

Sorbents are mostly used in the form of granules, usually of 44 BS mesh, or lumpy dust particles in perforated metal containers of one or two standard sizes. The standards refer in the first place to the weight of the sorbent, e.g., 0·25 and 0·5 kg ($\frac{1}{2}$ lb and 1 lb) tins. They are arranged in parallel to form batteries which can be used either in parallel or in series, the former for large air volumes, the latter for extreme purification.

Gross particulate (solid or liquid) pollution should be removed by impingement (impact) filters, or by electrostatic precipitation before the remaining molecular pollutant is admitted to the activated carbon beds.

As a general recommendation carbon is useful if the pollutant is in a concentration below 5 ppm. Higher concentrations require compound methods involving more than one type of odour elimination.

An instance is given by the Air Pollution Control Association's TA-7 Odour and Gas Treatment Committee, Technical Data Sheet No. 1 from which the description is extracted (136).

Lemon pulp is extracted with isopropyl alcohol involving an exhaust of 3350 m^3/h air at 50°C polluted with 160 l/h of the solvent vapour. The alcohol vapour first enters a wet-scrubber where it is stripped of particulates (lemon fibres) and of water solubles (acid fumes), and then passes through the activated carbon system where the isopropyl alcohol vapours are persorbed, and the purified air is released to atmosphere, containing less than 20 g of isopropyl alcohol (an osmogen) per hour. There are two tanks filled with activated carbon, the one persorbing the organic vapours from the air stream while the other tank, containing saturated carbon, is being steam stripped. Figure 26. The stripped alcohol vapours are condensed and pass into a storage tank whence they flow through a fractionating column yielding completely reusable solvent. The economics of this system are an essential item for consideration since the price of recovered isopropyl alcohol is but 14–15% of the purchase price of the solvent.

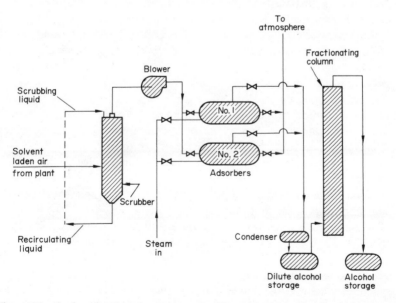

Figure 26   A plant for the adsorption, on activated carbon, of an organic solvent (136).

This saving includes all operational expense, i.e., electrical power, cooling water, steam, chemicals, maintenance, and supervision of the completely automatically working plant (328, 329).

The useful life of activated carbon is given by the approximate formula

$$t = 6\cdot4 \times 10^6 \times (r/a)\ (W/MVc)$$

where t = useful life in hours, r = fractional retentivity ($r = 0\cdot14 - 0\cdot5$), a = fractional persorptivity ($a = 0\cdot25 - 0\cdot8$), W = weight of the activated carbon (lb), M = molecular weight of the pollutant (weighted average, if several), V = volume of air passing through (ft^3/min), c = concentration (ppm) of the pollutant (weighted average, if several).

The values of fractional retentivity are 0·14 for acrolein type (burning of hot fat) pollutants, 0·36 for body and similar odours (animal odours, butyric acid, valeric acid), the highest values being characteristic of bromine, iodine, and some of their compounds, e.g., chloroform, iodoform, also of carbon tetrachloride, see Figure 27.

Whilst persorption is used as a most effective means of removing pollutants from an air stream, adsorption is a physical property which frequently causes an odour nuisance. Adsorption has been defined as the retention of a sorbate (gas or vapour) on the exterior surfaces of the sorbent. The sorbent in these cases may be the outer walls of a building, the total surface area of the leaves of a tree, the (considerable) outer surfaces of textiles, i.e., the outer surface of each fibre woven into the material, the microsurface structure of paper, for instance, of wallpapers especially of the thick, spongy type, and many more.

The forces bonding molecules to surfaces are much stronger than any mechanical forces applied for their removal, be they physical forces (vacuum cleaning) or chemical forces (washing down). Direct sunshine will energize some molecules sufficient for them to be released. When the entire outer surface of the sorbent is covered with molecules, the adsorbing action ceases. LANGMUIR has proposed that adsorbed layers are monomolecular similar to persorbed layers, but it seems that another layer may build-up on top of the first one.

The only means of controlling an odour nuisance caused by adsorbed molecules whether in the open (walls, leaves, grass, soil), or indoors (curtains, upholstery, carpets, wallpaper) is to oxidize the osmogenic matter. Indoors, this can be done quite conveniently by means of nascent oxygen which is produced in situ along, and at the target of, ultra-violet rays of wavelengths shorter than 2000 Å (cf. section 39). Out-of-doors, nascent oxygen is neither available naturally, nor can it be produced

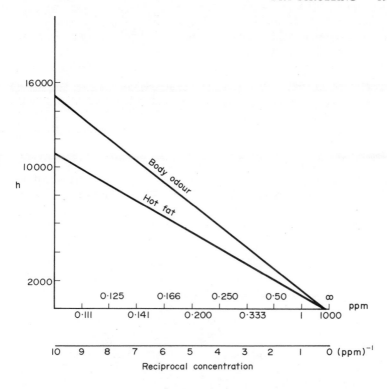

Figure 27   Useful life of activated carbon.

artificially at an economic figure for this purpose if the source of odour is extended as in the case of walls, etc. Hence, only sunshine may assist by photochemical transformations.

A new principle has been applied to dry-scrubbing using another gaseous phase as the separating medium. This is known as the helical flow technique (148, 222, 223). Originally evolved by SIEMENS, the dry-scrubber works by introducing the dustladen air axially into an upright cylinder by passing it across a guide cone or guide vane. This causes the polluted air to move in a twisted path so that strong centrifugal forces act upon the particulate matter in the air stream. The particles are thrown towards the wall of the cylinder and would come to rest there, were it not for a secondary air current which sweeps the cylinder wall in a helical downward movement collecting, and carrying the particulates into the receptor tank provided at the bottom of the cylinder. The drag on the secondary or sweeping air current is directed away from the cylinder wall so that any particle thrown towards it will fly in this direction, but will be diverted before actually reaching it and taken by the secondary air current into the collecting tank.

Removal efficiency for particle sizes above 1·5 μm is better than 90%, and above 4 μm better than 99%. Using hot secondary air allows of drying moisture-laden air or, vice-versa, cooling a hot polluted air stream if cold air is used. By manipulating the direction of the secondary air inlet relative to the radial cross-section of the cylinder it is possible to collect the pollutant into a ring which is kept rotating, yet stationary with regard to its position along the axis of the cylinder, Figure 28; or by slightly diverging the angle of entry from the previous value, the annular accumulation of fallout (dust or droplets) will form into a rotating and slowly progressing

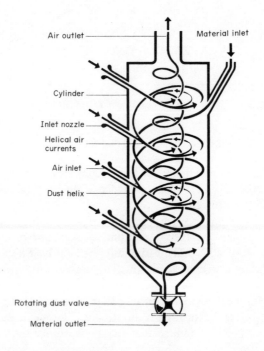

Figure 28   A drier using the helical flow technique. Helical air currents carry the dust along to the dust collector. (*Courtesy Siemens A.G.*)

helix. The direction of progress is towards the receptor tank. By this means a given quantity of pollutant may be kept within the cylinder for a pre-determined time. The residence or dwell time is controllable by the angle of the secondary air inlet, Figure 29.

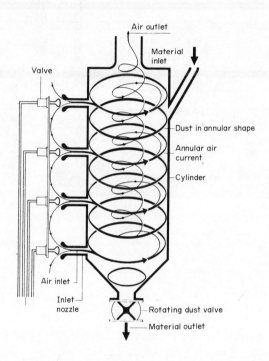

Figure 29   A drier using the helical flow technique. Annular stationary air currents provide the required residence time. (*Courtesy Siemens A.G.*)

Materials for scrubbers and ducting may be either steel, stainless steel or, depending on the danger of corrosion, the more resistant steels either with plastic coating or without. Polypropylene and polyvinylchloride offer a great scope, especially when reinforced with glass fibre. Stacks up to 100 m (300 ft) high, and packed towers have been built from glass re-inforced plastic with operating temperatures up to 100°C, Figure 30. Plastic Raschig rings of polypropylene often replace other materials used in packed towers (158).

Dry-scrubbing is an essential phase in the production of kiln-dried materials, such as pet foods, since large amounts of odoriferous dust are

Figure 30 Celmar extraction system installed at Courtaulds Ltd., Preston, England. Celmar is a synthetic material incorporating polypropylene and glass fibres. The duct operates under a slight negative pressure carrying waste gases, from viscose rayon spinning machines, containing hydrogen sulphide, carbon disulphide, and sulphuric acid mist. (*Courtesy British Celanese Ltd.*)

produced and carried together with the granular end product. Escaping dust produces an often inaccessible source of penetrating odour. Inaccessible, because the dust clings to structural building elements under the roof, is high upon the walls, and fills every nook and cranny between girders, roof trusses, and bricks. The more effectively dust is removed, the less trouble will be experienced with osmogenic pollution of large volumes of air.

34 Chemical cleaning

There are cases of air pollution which are not tractable either by wet or by dry phase separation. We must then consider chemical reactions by which to remove the pollutant from, or change its offensive constitution in, the air stream. As with all methods relying on additives, the cost of the additive is an item which over long periods, might well be prohibitive. In industrial applications such chemical reactions will be performed as yield a useful or saleable byproduct.

British Patent No 826 221 (1956) describes the removal of sulphur dioxide from stack gases by introducing gaseous ammonia into the stack at temperatures above dew-point. Ammonia salts are forming in the stack and are carried to an electrostatic precipitator making a readily saleable byproduct.

Ammonia in the gaseous state can be reacted with by an acid. Weak (organic) acids are preferable to strong (inorganic) acids. The reaction

$$NH_3 + CH_3COOH \rightarrow CH_3COONH_4$$
ammonia + acetic acid → ammonium acetate

is preferable to

$$NH_3 + HCl \rightarrow NH_4Cl$$
ammonia + hydrochloric acid → ammonium chloride

With simple, i.e., non-catalysed chemical reactions like the one above, a certain period of time must be allowed for the interaction of chemicals. On the average, 25–45 sec will be needed and this time lapse must be built into the ventilating system by designing extension chambers in the ducting system, or by injecting the reagent at a point sufficient distant from the air release point.

Where simple chemical reactions are insufficient, catalysis is resorted to. Catalysts are highly reactive media and for each type of odoriferous gas

199

a different catalyst is used. Catalysts do not take part in the chemical reaction sponsored by them, and the ratio of reacting chemical-to-catalyst is high, often $1:10^6$ to $1:10^8$. These are physical catalysts. They may take many forms, be chemicals themselves or, for instance, mechanical energy, such as air vibrations (ultrasound). Even audible, yet very high pitched, sound may accelerate a chemical process. This audible noise—at 13·5 kHz ($= 13\cdot5$ kc/sec)—has been fed into a heated nickel filament past which a gas stream moved (141). Using a very small sound intensity—only $0\cdot05$ W/cm^2—the decomposition rate of formic acid vapour increased by one-half, that of ammonia by one-sixth, whereas others did not improve at all. Hydrogenation of ethylene C_2H_4 to ethane C_2H_6 took place, owing to the nickel catalyst, at the usual rate, which could not be improved by insonating the gas. Sound or ultrasound will precipitate solid and liquid particulates from the air, improving the reaction rate (either in speed or quantity) of a physical catalyst. Ultrasonic sirens can deal with air volumes up to 100 m^3/min (142).

There are chemical catalysts, i.e., catalysts which take part in the chemical reaction, but are re-formed in the last stage of the process. Such a process is described in British Patent No 803 801 (1958) concerning the removal of hydrogen sulphide from the exhaust air of nylon manufacturing plant. The volumes of air are large, the concentration of hydrogen sulphide small. The proposed process removes the offending pollutant in a single stage using ferric hydroxide as a catalyst.

$$6H_2S + 4Fe(OH)_3 \rightarrow 2Fe_2S_3 + 12H_2O$$
$$2Fe_2S_3 + 3O_2 + 6H_2O \rightarrow 4Fe(OH)_3 + 3S_2$$

The required ratio of catalyst to reagent is 30 molecules of ferric hydroxide for each molecule of hydrogen sulphide.

A forerunner of this process uses ferric oxide, Fe_2O_3, and is a multistage process. Sodium carbonate in aqueous solution is made to react with hydrogen sulphide.

(1) $6Na_2CO_3 + 6H_2S \rightarrow 6NaHS + 6NaHCO_3$
                       sodium   sodium bicarbonate
                   hydrosulphide

(2) $6NaHS + 2Fe_2O_3 \rightarrow 2Fe_2S_3 + 6NaOH$

(3) $2Fe_2S_3 + 3O_2 \rightarrow 2Fe_2O_3 + 3S_2$

In both instances decomposition of the catalyst occurs and its subsequent reconstitution leaves elemental sulphur. The ferric hydroxide process takes place in air, the ferric oxide process in water.

The magnitude of the problem in the synthetic fibres industry is quite gigantic. In the production of viscose staple fibres total air flow may be 50 000 m^3/min, and more. The air may carry 300–500 ppm of hydrogen sulphide, and about three times that amount of carbon disulphide. Other osmogens may also be present, but in much smaller proportions. There may be droplets of sulphuric acid and of sodium sulphate entrained in the air stream. It is essential to provide sufficient oxygen (air) as, otherwise, the ferric sulphide will split into ferrous sulphide and iron disulphide

$$Fe_2S_3 \rightarrow FeS + FeS_2$$

The ferric sulphide process removes some 50–70 ppm carbon disulphide and 280–300 ppm of hydrogen sulphide.

Yet another process is described in British Patent No 838 571. Trays in a tower are packed with ferric hydroxide $Fe(OH)_3$ in an alkaline suspension. Contact time required is about 20 sec and hydrogen sulphide removal is 20 g for each 100 g of the ferric hydroxide catalyst. This is a conversion ratio of 100:20 or 5:1.

Hydrogen sulphide is one of, if not the most, offensive of industrially produced obnoxious odours, and many patents or processes are concerned with its removal and destruction, i.e., chemical or physical transformation into a non-smelling compound. Hydrogen sulphide in quantity is also produced by thermal power stations, furnaces, by the petroleum industry, and others. Of great promise is, therefore, the use of chelated iron compounds as described in British Patent No 855 421 (1960), and the Stretford Process devised by the National Coal Board in co-operation with the Clayton Aniline Company. A chelate (pronounce kélate) is a molecular structure in which a ring is formed by the residual unshared electrons (valences) of neighbouring atoms. The word is derived from the Greek khēlē = claw. The catalyst may be a complex form of iron with an amino acid. In the presence of hydrogen sulphide the compound is reduced to ferrous iron. By aerating the ferrous iron, FeO, it will be oxidized and return to the ferric form, Fe_2O_3. Elemental sulphur is formed in the end and separated by filtration or settling out (137).

The Stretford process (developed at the Stretford Works, near Manchester) uses a liquor of various sodium salts with anthraquinone disulphonic acid, $C_{14}H_8O_8S_2$. The polluted air and the chemically doped washing solution move counter-currentwise in a packed tower. The pregnant solution is then treated with carbonyl radicals and produces elemental sulphur. Complete oxidation of the absorbed hydrogen sulphide occurs only if the mixutre of 2,6-anthraquinone disulphonic acid and 2,7-anthraquinone disulphonic acid has a pH value of 8·5–9·5, and if the acid is available in excess of molecular requirements. Although the Stretford process was originally developed for treating waste air from gas works, it is now more widely applied in industry for treating polluted streams of gases of different composition (138).

The HAINES process (139), another method employing a chemically doped liquid phase, uses a synthetic hydrated aluminium sodium silicate (zeolite) of the general formula $Na_2Al_2H_6Si_2O_8$. The zeolite first adsorbs hydrogen sulphide, but is then regenerated with sulphur dioxide which combines to form elemental sulphur

$$4H_2S + 2SO_2 \rightarrow 4H_2O + 3S_2$$

The capacity of the process is on a large scale. In practical applications 2 million m^3/day of air polluted with 15 % hydrogen sulphide yield 40 t of liquid sulphur per day.

Manganese is used in the catalytic formation of dilute sulphuric acid from sulphur dioxide and oxygen (air). Another group of catalysts are the nitrated phenols, especially trinitrophenol, $C_6H_3O_7N_3$, which directly oxidize hydrogen sulphide according to

$$2H_2S + O_2 \rightarrow 2H_2O + S_2$$

This is a process developed by the Osaka Gas Company and is simple to operate. The gas, containing hydrogen sulphide, is mixed with air and blown across the nitrated phenol in solution whereupon elemental sulphur forms as a sludge which is skimmed off or separated by filtration.

Sulphur dioxide removal is generally more difficult than that of hydrogen sulphide. One of the most economical processes has been devised by GOUCHARENKO (140). Sulphur dioxide is hydrolized yielding sulphurous acid

$$SO_2 + H_2O \rightarrow H_2SO_3$$

Washing with calcium hydroxide

$$H_2SO_3 + Ca(OH)_2 \rightarrow CaSO_3 + 2H_2O$$

or with calcium carbonate

$$H_2SO_3 + CaCO_3 \rightarrow CaSO_3 + H_2O + CO_2$$

yields calcium sulphite which is soluble in sulphurous acid

$$H_2SO_3 + CaSO_3 \rightarrow Ca(HSO_3)_2$$

producing calcium bisulphite which is water soluble.

Sulphur dioxide is converted to sulphuric acid in a two-step process, the first step being rapid adsorption followed by the slow chemical conversion into the acid. The rate of oxidation depends upon the concentrations of the pollutant, of water vapour and oxygen (air), on the temperature of the waste gas and its residence time in the reactor bed. Pore size and pore distribution in the adsorptive reagent (coke) also determine the rate of conversion. Saturated coke is thermally regenerated in a stream of inert gas, mostly nitrogen (144).

Quaternary ammonium (NH_4OH in which all four hydrogen atoms are replaced by a radical, such as ethyl, C_2H_5) is used as a denaturing agent for exhausts from internal combustion engines, US Pat. 2 932 364; 1960.

Iodine is an effective chemical catalyst specifically suitable for the removal of aldehydes from air. The iodine enters into a chemical bond with the aldehyde (step one), and immediately recombines by reacting with the intermediate iodo-compound which has formed with the appropriate radical in the methane series (step two). Taking acetaldehyde, $CH_3.CHO$, as an instance

(1) $CH_3.CHO + I_2 \rightarrow CH_3I + HI + CO$
    acetaldehyde

(2) $CH_3I + HI \rightarrow CH_4 + I_2$
    iodomethane   methane
          hydrogen iodide

or, generally,

(1) $C_nH_{2n+1}.CHO + I_2 \rightarrow C_nH_{2n+1}.I + HI + CO$
(2) $C_nH_{2n+1}I + HI \rightarrow C_nH_{2n+2} + I_2$

Gaseous phenols (hospital smells) are absorbed by impact scrubbing with 0·1 N caustic soda solution (143).

A large and important section in chemical washing is taken up by those methods which cause the pollutant to become oxidized, i.e., inoffensive. Though it is possible, and mostly simpler and more economical, to use physical means, i.e., ultra-violet radiation for oxidizing osmogenic matter, the chemical method may sometimes be worth considering as an alternative. There is a variety of substances the molecules of which contain a surplus of oxygen atoms which split off easily, producing nascent oxygen in situ. Two of the most likely are potassium permanganate, $KMnO_4$, and chlorine peroxide, ClO_2. (In the presence of ammonia, NH_3, hydrogen sulphide, H_2S, hydrogen phosphide, H_3P, or methane, CH_4, chlorine peroxide *must not* be used, because the combination would form an explosive mixture, as it does with sulphur contained in rubber, and with sodium chlorite, $NaClO_2$.)

Nascent oxygen as produced from chlorine reactions or from air cannot deal with such pollutants as methyl or ethylamines, sulphydryls, and sulphides, but chlorine peroxide can oxidize these osmogens.

Chlorine, Cl	Chlorine peroxide, ClO_2
does react with	directly oxidizes
amines sulphydryls sulphides	
is effective only in an acid medium	is effective in alkaline, acid, or neutral media
in water produces intermediates which often have an objectionable smell	oxidizes directly without first chlorinating the water or organic vapours
does not oxidize in the presence of ammonia, but forms chloramine and ammonium chloride	oxidizes in the presence of ammonia
is dispensed from steel cylinders	is generated in situ from sodium chlorate, sulphuric acid, and methanol

Chlorine peroxide as well as chlorine gas corrode steel, aluminium, and other metals, but glass, ceramics, asphalt, bitumen, several plastics, and synthetic rubbers, are corrosion resistant.

Chlorine peroxide boils at 9·9°C, and explodes at 100°C, as well as in contact with certain other substances. It is soluble in cold water, 2000 cm^3 of ClO_2 in 100 ml of water at 4°C; it is soluble in alkali and sulphuric acid.

35 Dilution

Air deodorization by dilution can mean either of two things:

(1) the dilution of polluted air which is, generally, carried out by discharging—often direct and without preparatory treatment—at a point high enough above the ground to get the osmogenic particles into fast moving natural air currents, thereby reducing the concentration to a tolerable degree. The point which must not be forgotten is, of course, the probability of reconcentration of the dispersed osmogens on surfaces of walls, leaves, etc.;

(2) the dilution of the scrubbing water which, in the process of washing the polluted air, has become a source of smell itself, giving off fumes or causing chemical reactions in the sewers producing obnoxious odours.

Anti-pollution treatment of air by dilution is not good practice and cannot be recommended although it is a favourite method adopted by those whose premises are not too close to residential districts, or lie favourably with regard to prevailing winds. A measure of justification can be accorded to this method if the works stand on a hill high above habitation, and if the chimney is tall enough to guarantee a high exit velocity of the fumes. These will then reach the air strata of high wind speeds, and dispersion is satisfactory. The works must not, however, overlook a valley, because of the danger of downwash.

Where natural draught in a chimney is insufficient for this purpose, fans should be installed.

Ventilation or exhaustion pure and simple is useful as a means of changing the air when the cause of the smell is purely domestic, or occurs sporadically and then only for short durations, and of low intensity. Opening the windows is a system of natural exhaustion as good as any, but should not be permitted as the only means of ventilation in e.g. a waste processor's plant where steam digesters are discharging—though the opinions of the works manager and the Public Health Inspector differ fundamentally on this point. It should also be considered that even small and, apparently, innocuous quantities of smell-polluted air may produce cumulative effects under adverse weather conditions, such as humidity, temperature, and low barometric pressure, and that local climate plays a decisive role in the movement of air which, often, results in static smells 'hanging about the place'.

The means of achieving dilution of odour-polluted air in free atmosphere is its release through a properly designed stack or chimney. Chimneys are not primarily built for the purpose of discharging odorous matter: this is incidental to their task of creating the draught necessary for maintaining the fire and taking away the waste of combustion. A characteristic of a chimney is its height, and the shape the dispersed matter will take under the influence of the winds once it has been discharged into free air.

The mouth of the chimney is considered a fixed elevated point source with definite characteristic data. These are lacking in an object on ground level such as a factory, because of the uncontrolled emission of polluted air from the large, three-dimensional object. For the purpose of evaluating emissions from fixed elevated point sources, chimneys, smoke stacks, vent shafts, and similar structures discharging polluted air or gas with or without white steam may be considered under the same heading.

The mouth of the chimney can be regarded as the origin of a rectangular coordinate system, the y-axis coinciding with the vertical axis of the chimney and representing the velocity of the mean cross-sectional updraught in the structure, while the x-axis is horizontal and represents the atmospheric air movement or its horizontal component moving at velocity c. Once the air has left the mouth of the chimney, its velocity v is inversely proportional to the distance from the chimney mouth. The product of volume and velocity is constant for undisturbed emission. A wind blowing across the chimney top at velocity c will deviate the plume of gases from the path of natural buoyancy and divert it in the shape of a cone in the direction of the wind. The apex of the cone may be assumed at the centre of the chimney mouth, i.e., at the origin of the coordinate system, Figure 31. The angle α is a function of both the speed v of updraught, and the wind speed, c, thus $\tan \alpha = v/c$. From Figure 31, it can be seen that $\tan \alpha = H/D$, where H is the height of the chimney, and D the distance from the chimney for which $r = H$, i.e., for which the radius, r, of the cone of polluted air equals chimney height. For an observer at P, for which $r = H$, this will be the point nearest to the chimney where the plume hits the ground and the smell of the discharge from the chimney mouth will be perceived. For values of $r < H$ and $D' < D$ there is an

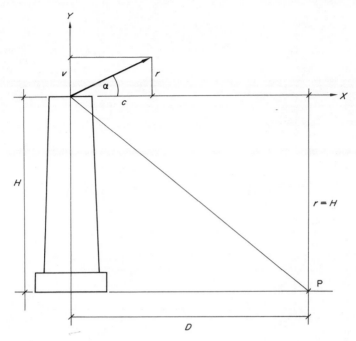

Figure 31   Stack exit velocity v and wind velocity c shape the plume emanating from the chimney.

odourfree zone on the ground in the windward direction beginning at the base of the chimney and extending to its maximum distance for which $r = H$, and $D' = D$.

For small values of c/v ($= \cot \alpha$) the axis of the cone will not be in the x-direction, i.e., horizontal. For large values of c/v the cone axis will be nearly horizontal, and the coefficient of dilution is almost independent of the direction of flow, Figure 32. The coefficient is generally taken to be constant at 13 %.

SUTTON's formulae are semi-empirical and are used for calculating the concentrations of pollutants in plumes emitted from an elevated point source (26). They have been reviewed and extended several times in an effort to obtain a more realistic evaluation of the diffusion over large distances. PASQUILL (23) has notably contributed to this subject. The concentration at any point in the plume emitted from the mouth of a chimney of given height is expressed by a Gaussian formula. Graphs and tables show the relationship of the diffusion coefficients in a free atmosphere as a function of distance from the point source, $y = f_1(x)$, $z = f_2(x)$.

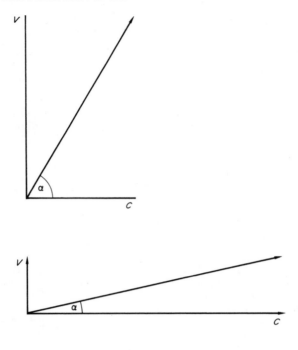

Figure 32   Stack exit velocity v and wind velocity c determine the rise angle of the plume.

Another quantity profoundly affecting the coefficients is atmospheric instability which affects the distribution of pollutants within the plume. Examples, calculated for various metereological conditions and chimney heights of 20, 30, 50, and 70 m have been published by BRANCA (24). PASQUILL'S approach to the problem has advanced it considerably, though his premises exert rather severe limitations on their general applicability. PASQUILL'S equations are valid only in the case of the source (factory and chimney) standing isolated on flat grassy land. Tall buildings or trees which affect the formation of the plume and the distribution of pollutants in it by causing cross draughts and random air currents, are not taken care of by his formulation. Great experience, a good knowledge of all technical data as well as a full appreciation of geographical, topographical, and microclimatic data is essential for a successful solution.

BRUMMAGE (282) has come to the conclusion that SUTTON'S formula for the dispersion of flue gases from a single stack is sufficiently accurate.

For stacks producing gaseous effluents in the range 15–100 m³/sec (at standard temperature and pressure) the formula for plume rise best suited to these conditions is

$$\Delta h = KQ_h^{0\cdot5}/c^{0\cdot75}$$

where Δh = plume rise, Q_h = heat flux, and c = wind velocity at the height of the top of the stack.

This equation was incorporated by BRUMMAGE into SUTTON's formula giving the critical ground-level concentration

$$C_{crit} = 6\cdot30 \times 10^2 \times Q_p \left(\frac{1}{Q_v \Delta Th}\right)^{\frac{2}{3}}$$

in which the concentration C_{crit} is expressed in mg/m³ at half-hourly samples, Q_p is the production rate of the pollutant, (in this case SO_2) in kg/h, Q_v is the rate of production of total flue gas in m³/h (at standard temperature and pressure), and ΔT is the temperature difference between the flue gas and the surrounding air in °K; h is the height of the stack in metres.

The validity of the exponents used by BRUMMAGE and others has been challenged by STONE and CLARKE (in 27) of the Planning Department, Central Electricity Generating Board, London, who have shown that thermal rise continues for a minimum of 1000 m downwind of a tall stack. They find ample experimental evidence for the exponents used in the Brummage formula to be 0·25 and 1·00 respectively, making the plume rise

$$\Delta h = kQ_h^{0\cdot25}/c.$$

BOSANQUET (25) has calculated tables for the computation of the rise of hot gases when the exit velocity, v, at the mouth of the chimney is either nil, or is not negligible. A graphical solution of the BOSANQUET–PEARSON equation for maximum concentration on the ground at various distances from the source, and for an emission rate of 0·3 m³/min ($=10$ ft³/min) at normal atmospheric temperature and a wind velocity of $c = 30\cdot5$ m/min ($=100$ ft/min) is shown in Fig. 33. The diffusion coefficients p and q vary with atmospheric instability, but the ratio of p/q remains near enough constant.

Turbulence	p	q	p/q
High	0·10	0·16	0·63
Average	0·05	0·08	0·63
Low	0·02	0·04	0·50

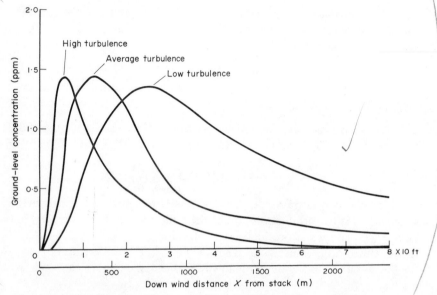

Figure 33   Effect of atmospheric instability on diffusion from a chimney.
Data: Stack height 100 ft = 31 m

    Emission rate 10 ft³/min = 0·3 m³/min

    Wind speed 100 ft/min = 31 m/min

(*From* Manufacturing Chemists' Association Inc. *Air Pollution Abatement Manual*, Chapter 8)

The effect of atmospheric instability on the diffusion of pollutants emitted from a stack is illustrated by the following figures.

Turbulence	Peak concentration at ground level ... m from source	Concentration, ppm, at a distance of 1000 m	2000 m
High	240	0·2	0·05
Average	400	0·5	0·15
Low	750	1·2	0·55

BLOKKER (114) has evolved a method for calculating dispersion, plume rise, and maximum concentration at ground level of a gas emanating from a stack. He gives a list of horizontal and vertical dispersion parameters as a function of atmospheric stability which are then used in SUTTON's dispersion equation. BLOKKER arrives at, principally, the same relationship as BRUMMAGE, and also suggests various factors which, depending on atmospheric conditions, allow one to calculate chimney height to a fair degree of accuracy.

These considerations are valid for single stacks. When multiflue stacks are used, the flues should be designed to project over the top of the chimney to prevent downwash of the plume, Figures 34a and 34b.

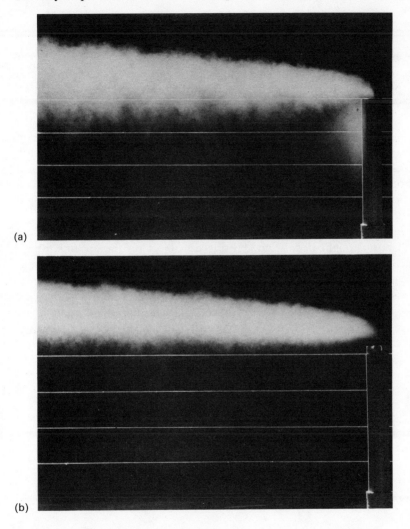

(a)

(b)

Figure 34   Wind tunnel test on multi-flue stack.
(a) Downwash occurs with flush-topped design.
(b) Improvement in emission with projecting flues.
In both tests, steam has been used in order to trace the behaviour of the plume as it left the chimney. (*Courtesy Central Electricity Generating Board* (27))

Japanese researchers have put forward since 1952 (INOUE), again in 1959 (OGURA), and in 1966 (HINO et al.) that the relationship between sampling time, t, and mean ground level concentration, C, should be a $-1/2$ power law,

$$C = At(\exp -1/2)$$

where A is a proportionality factor. The concentration of the pollutant decreases obviously with increasing sampling time, because the lateral dispersion of the effluent increases with time. HINO reported (209) that experimental data give strong support to the $-1/2$ power hypothesis.

Peak concentration at ground level occurs nearer to the source, the higher the atmospheric turbulence, and conversely. This is an important finding resulting in a new approach to the problem by designing the tall stack. Primary considerations are of a meteorological and topographical nature. These are the important questions: where is the source of pollution situated, and how does the air flow in consequence of surface configuration?

In Figure 35, air currents higher than about 200 m above flat ground will flow straight downwind. Hitting an object on the ground the air current will eddy round it, and make contact with the ground at various distances from the intercepting object, be it a tall building, a row of poplar trees, or a hill side. If a hill lies in the path of the chimney plume, the downwash on the leeward side of the hill will deposit odorous matter in an area which might have been believed safe from the incidence of airborne smell.

The mouth of any chimney should be above the straight air flow pattern, here assumed to be at 200 m.

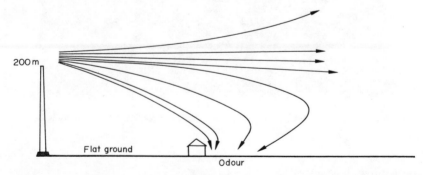

Figure 35 The lower strata of a plume have the tendency to cause an odour nuisance even at a considerable distance from the stack. The upper strata are taken up by the winds.

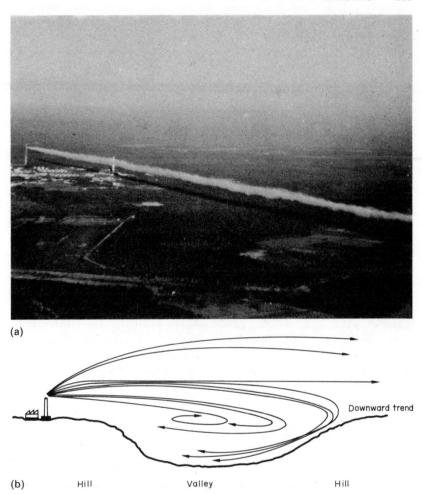

(a)

(b)   Hill   Valley   Hill

Figure 36   Winds fanning out over a ridge or hill and across a valley come down
into the valley after hitting the opposite side of the hill. Chimneys on the windward
hill must be high and well away from the valley to prevent a stagnant pool of
polluted air hanging about in the valley.

If the ground has a deep depression, e.g., a valley, Figure 36a, winds
blowing across the top of the valley and close to it, will hit the opposite
side, or be depressed, or drawn, into the valley towards its bottom. The
overall effect is an airflow in a circular pattern, the circles lying in vertical
planes, Figure 36b. An air current moving across a valley at height will not
be affected by the ground depression, and will bridge the valley effectively.

Polluting matter must clearly never be allowed to get into the air currents which may circle round in a valley.

British and American practice is discussed in a review issued by the National Coal Policy Conference (27). Various aspects of air flow patterns in and over valleys relevant to air pollution are discussed by HEWSON (33) and DAVIDSON (34).

Stacks of different heights, diameters, and collecting efficiencies have been investigated to find the optimum design data. These are then considered from a cost point of view.

Wind tunnel experiments are used more frequently now than before in order to determine the most suitable orientation of buildings and location of stacks so that no eddies are produced which would carry air pollution back to ground level (28). The dispersion of polluting matter can be influenced by raising the height of the chimney (29). Short stacks, though cheaper to build, are unreliable to operate when the removal of air pollution is imperative. The behaviour of plumes from short stacks— the term 'short' being defined by the prevailing atmospheric and meteorologic conditions with regard to wind direction and velocity as explained above—especially near buildings or trees, or nearby hills, or with gusts of wind, is complex and difficult to evaluate. Even then, there is no permanency in air movement. The effective configuration of the plume source can vary from that of a point source to an area source, and behave as either on ground or elevated level. The change-over from the one to the other condition may just take seconds (30).

Odour intensity at ground level is fundamentally affected by meteorological conditions, especially by an inversion. Graphically, an inversion is characterized by an inverted temperature lapse, Figure 37, which means that temperature increases instead of decreasing, with height. A uniform inversion will compress the plume so that its contents cannot be dispersed. The cone is now long drawn and narrow. As the narrow cone may not touch the ground at all or, only at much greater distances than normally, odour intensity at ground level may be reduced by a factor between 5 and 10.

Atmospherics do not always produce either a normal or an inverted lapse rate. Conditions may be such that an inversion overlays a layer with normal lapse rate in which case the upper longitudinal part of the normal cone is depressed by the inversion, or the layer with a normal lapse rate rests on top of the inversion 'bubble' in which case the lower part of the plume is severely curtailed, and does not penetrate the ceiling of the 'bubble' at all. This means that the discharge from the chimney will not reach the ground, which remains free from pollution. Inversions occur at

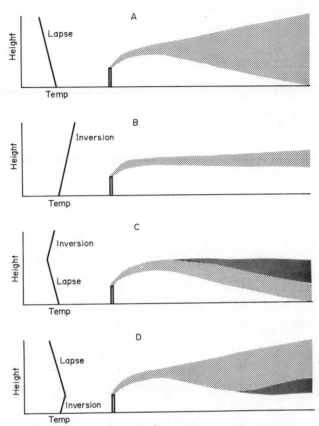

Figure 37   Diagrammatic representation of plume dispersion under different conditions of atmospheric stability. (*Courtesy Central Electricity Generating Board* (27))

any height though those at lower altitudes are most frequent. A chimney discharging in a 'bubble' will dangerously pollute the air in it.

The effects of meteorological and climatic conditions on air pollution are best illustrated by the magnitude of the 'Los Angeles problem'. It is the largest heavily-industrialized semitropical area in the world. Average wind speeds are only 10 km/h which means that one hardly feels the air moving over the face, and temperature inversions occur on more than 260 days in every year. They severely restrict dispersion of air pollutants produced by the activities of some 7 million people living in the 'bubble' for more than two-thirds of their life. Industrial establishments have nearly quadrupled in number since 1939, and the population has more than doubled.

The last detail sketch in Figure 37 suggests a ready solution to the problem. By designing chimneys of such a height that most of the lower inversion ceilings remain below their mouths, pollution at ground level will hardly be noticeable. Tall chimneys are really tall compared with usual designs. The American Electric Power System has at Maudsville a chimney 400 m high (113). Inversions at high level (Figure 38), which are very frequent, have been discussed by SCORER (242).

Figure 38   Plumes from the 320 ft (nearly 100 m) high stacks at Deptford Power Station, London, dispersing above the fog surface, 750/800 ft (about 250 m) above ground level. The presence of fog was an essential part of the experiment. By saturating the plume with water it rose through the temperature inversion and can be clearly seen above it. (*Courtesy Central Electricity Generating Board* (27))

The basic recommendations about chimney heights and air pollution for coal or oil fuelled plants must take into account both soot and sulphur dioxide emission. Soot and other particulate emissions can be controlled by using the right type of fuel (desulphurized), the correct amount of combustion air and, as the final safeguard, electrostatic precipitation.

The same principles also apply, surprisingly, to vent-shafts and similar minor structures, even those not more than 10 m high. A brief case history will justify the point. A new residential district had been designed to take advantage of a hillside with a magnificent view. Row upon row of bungalows were built, Italian fashion, i.e., one above the other. To vent the sewers, shafts were provided at the correct points. After the people had bought their houses and moved in, and lived there for a short while, letters reached the Local Authority complaining of a foul sewer smell which enveloped the whole district, and blew straight into open doors and windows. Investigation revealed that the trouble did not arise in the sewage works, but was caused by the ventshafts which had been erected as shown in Figure 39. Now, if these ventshafts had been chimneys, considerable thought would have gone into their siting, and the planners would have made sure that smoke from the houses below would not be blown by the prevailing wind into the houses above them. However, smells cannot be seen and their trails cannot be followed easily, so the possibility of foul sewer air pervading the houses never entered the mind of the designers.

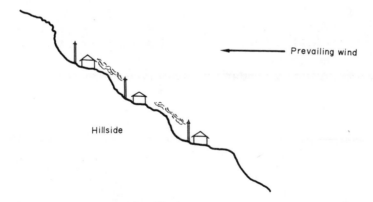

Figure 39   Houses of a newly developed hillside estate were so distributed in terraces that foul air from the septic tanks of one row of houses blew straight into doors and windows of the houses above.

Since the passing by the UK Parliament of the Clean Air Act (4), the Ministry of Housing and Local Government has assisted designers by issuing from time to time various Memoranda, advisory specifications and recommendations; notably *Chimney Heights* (Clean Air Act 1956), MHLG Circular No. 50/67 and *Chimney Heights*, second edition of the 1956 Clean Air Act Memorandum, both published by HMSO, London. The revision of the Memorandum reveals an advance in thinking. Whereas the first issue based its recommendations on the boiler rating, the second edition uses the criterion of weight of sulphur dioxide emitted per hour. For the engineer dealing with air pollution from chimneys the nomograms for calculating chimney height with reference to the osmogenic emission expressed in lb/h are of great interest.

The Memorandum specifies different chimney heights for different rates of emission.

Area classification

Class A   An underdeveloped area where development is unlikely, where background pollution is low, and where there is no development within half a mile of the new chimney;

Class B   A partially developed area with scattered houses, low background pollution, and no other comparable industrial emissions within a quarter of a mile of the new chimney;

Class C   A built-up residential area with only moderate background pollution and without other comparable industrial emissions;

Class D   An urban area of mixed industrial and residential development, with considerable background pollution, and with other comparable industrial emissions within a quarter of a mile of the new chimney;

Class E   A large city, or an urban area of mixed heavy industrial and dense residential development, with severe background pollution.

Rates of emission

Very small emissions	3– 30 lb/h	(1·3– 13 kg/h)
Small emissions	30– 100 lb/h	(13 – 45 kg/h)
Medium emissions	100– 400 lb/h	(45 –180 kg/h)
Large emissions	400–1800 lb/h	(180 –800 kg/h)

Range of chimney heights

Rate of emission	Range of chimney heights	
Very small	16– 72 ft	(5–21 m)
Small	40– 97 ft	(12–29 m)
Medium	70–160 ft	(20–45 m)
Large	100–220 ft	(30–60 m)

These heights apply immediately if the chimney is at least 1·5 times taller than the factory building, or nearby buildings. If the chimney height is less than 2·5 times the height of any buildings, a correction must be made to account for the reduction in effective height. The correction factor is determined by the dimensions of the building, Figures 40(a) to (e).

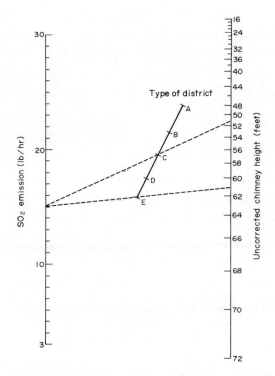

Figure 40(a) Uncorrected chimney heights—very small installations. (Nomogram from *Chimney Heights*, 2nd edition of the *1956 Clean Air Act Memorandum*, reproduced by permission of the Controller, HM Stationery Office)

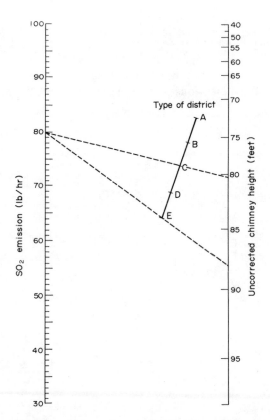

Figure 40(b)   Uncorrected chimney heights—small installations. (Nomogram from *Chimney Heights*, 2nd edition of the *1956 Clean Air Act Memorandum*, reproduced by permission of the Controller, HM Stationery Office)

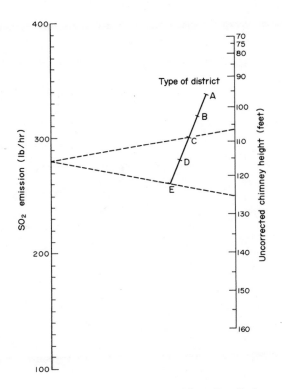

Figure 40(c)   Uncorrected chimney heights—medium installations. (Nomogram from *Chimney Heights*, 2nd edition of the *1956 Clean Air Act Memorandum*, reproduced by permission of the Controller, HM Stationery Office)

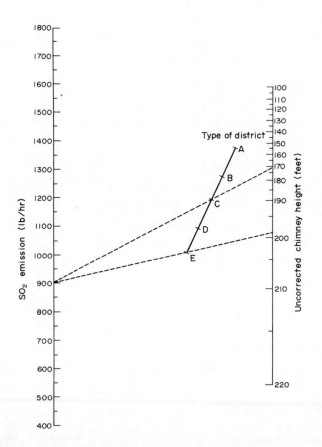

Figure 40(d)   Uncorrected chimney heights—large installations. (Nomogram from *Chimney Heights*, 2nd edition of the *1956 Clean Air Act Memorandum*, reproduced by permission of the Controller, HM Stationery Office)

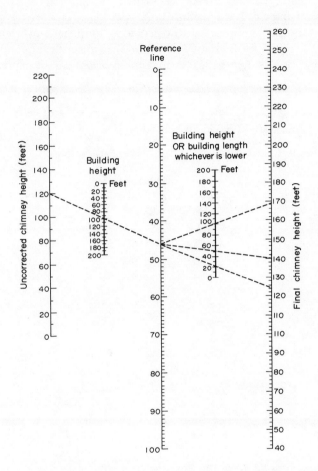

Figure 40(e)   Final chimney heights. (Nomogram from *Chimney Heights*, 2nd edition of the *1956 Clean Air Act Memorandum*, reproduced by permission of the Controller, HM Stationery Office)

A useful aid to the designer is the Brightside *Chimney Design Manual,* the 15 nomograms of which are based on the recommendations of the Ministry of Housing and Local Government. The Manual is published by the Research and Development Department of the Brightside Heating and Engineering Co. Ltd., Sheffield, England.

There are other approaches to calculating the height of a chimney. Adopting the velocity constant v/c ($= \tan \alpha$) as the critical quantity, and using BOSANQUET's theory (35) the maximum height can be expressed by

$$H_{\max} = \kappa(V/v)^{0.5}$$

where V = the volume flow of the waste gas, v = the exit velocity at the chimney mouth, κ = the height coefficient.

Wind velocities c are expressed on the Beaufort Scale.

Table 38   *Height coefficients*

v/c	κ
0·25	0·125
0·33	0·192
0·50	0·354
1	0·918
2	2·292
3	3·736
4	5·197

Table 39   *Beaufort Scale of wind velocities*

Beaufort	Effect	ft/sec	mi/h	m/sec
0	Smoke rises vertically	0	0	0
1	Smoke drifts; no sensation of air movement	3	2	1
2	Wind felt on face; leaves rustle	7	5	2
3	Leaves and small twigs in constant motion	15	10	5
4	Dust and loose paper raised	22	15	7
5	Small trees in leaf sway	31	21	9
6	Large branches move	41	28	12

The type and effect of meteorological parameters entering into the calculations of chimney heights are evaluated by KLUG (36) who also discusses the recommendations of the Verein Deutscher Ingenieure (VDI) concerning the minimum height of chimneys.

GATE (37) has shown that the magnitude of gaseous concentrations in air will be independent of the height of the chimney if the wind velocity is considerable, but that the intensity of smell is proportional to the inverse square of the distance from the chimney. He has suggested draught velocities of 10–12 m/sec, the latter being the maximum to avoid large air pressures in the chimney. Dense waste gases descend rapidly from the point of discharge, hence they should be emitted from the chimney mouth at the highest possible velocity so that they are carried into the strata where dispersion is no longer affected by buildings, trees, and surface configuration (38). A method of estimating the diffusion of windborne material has been described by PASQUILL (23) for a wide range of meteorological conditions and over distances up to 100 km (60 miles).

Wind velocities at various heights can be taken from a Table compiled in accordance with the formula recommended by the Meteorological Office. The windspeed c_H at any height H refers to the velocity measured at 10 m (33 ft) above ground level, thus

$$c_H = c_{10}(0\cdot1\,H)^{0\cdot17}$$

Table 40   *Wind speeds at various heights*

Height, m	Wind speed, m/sec							
10	3	6	9	12	15	18	21	24
20	3·4	6·8	10·2	13·6	17·0	20·4	23·8	27·2
30	3·6	7·2	10·9	14·4	18·1	21·8	25·3	28·9
45	3·9	7·8	11·7	15·5	19·4	23·3	27·2	31·1
60	4·1	8·2	12·2	16·3	20·4	24·4	28·5	32·6
75	4·3	8·5	12·8	17·0	21·3	25·5	29·8	34·0
100	4·4	8·7	13·1	17·4	21·8	26·2	30·5	34·9

Observations from the air have greatly assisted in following and outlining the smoke trails from chimneys and experimental installations. Air samples taken at various altitudes and at ground level are analyzed and yield information about dispersion of particulate and gaseous pollution.

The concentrations are then charted in horizontal as well as vertical planes producing a three dimensional flow diagram of the waste matter.

Uncontrolled discharges can yield concentrations of obnoxious, sometimes toxic, wastes especially from chemical works, greatly in excess even of the most tolerantly conceived permissible upper limits. As an instance, some of the data from the work by AVOYAN (31) is quoted. In addition to solid particulates, mostly carbides, carbon dust, and lime-stone, large amounts of cyanide compounds, nitrous oxides, and sulphur dioxide, were discharged from the chimney of a chemical factory. The area surrounding the plant was divided into circular zones with the chimney at the centre, redii of 500 m, 1000 m, 2000 m, and 3000 m. Dust was precipitated at 3–5 times the permissible concentration, even at the far perimeter of the area. Because of the specific gravity of sulphur dioxide (2·26 for air = 1·00) concentrations at ground level are much more serious than in the atmosphere. In the innermost zone concentration was 11 times, and at 1000 m distance 3 times, the permissible value. Similarly, nitrous oxide and cyanide compounds were far above the permissible concentrations. All measurements were made in the leeward direction.

The emission of hydrogen sulphide and carbon disulphide from a viscose plant was charted by JEDRZEJOWSKI (315). The rate of emission was 14·4 g H_2S and 42·2 g CS_2 per second for a daily output of 20 t viscose. Two series of measurements were made at wind velocities of 2·5 m/sec and 10 m/sec.

	Concentrations in mg/m^3			Permissible concentration
	Distances from the chimney			in mg/m^3
	2200 m	2400 m	3000 m	
CS_2	0·082	0·138	0·220	2·3
H_2S	0·028	0·047	0·075	0·0015

For observations of flow patterns in the chimney stack, windows are built into the brick or concrete work and platforms are put in convenient positions for the investigator. Soot, tar droplets, fly ash, and any other particulate matter carried in the waste gases will soon blot out the window so that direct visual inspection, or the use of photoelectric or television equipment will suffer from inaccuracies and lack of detail. There is a design available for chimney inspection windows which require attention about only once a year (39), and remain clear and serviceable for the whole of that time.

In tests with plumes in free air the optical properties and visual effects are of primary importance. If, as in the case of osmogenic gases and vapours, their movement cannot be observed directly because of their transparency to light, artificial smokes are released with them simultaneously. CONNER and HODGKINSON (40) have studied this problem. Results indicate that visual effects vary with the background of the plumes, with illumination and viewing conditions. Variations in results also occur with different shades of the plume, being greater for white plumes than for black ones. The amount of aerosol suspended in the plume is best evaluated optically by its transmittance. Methods of measuring transmittance are given.

'Natural' ventilation is inadequate particularly when kitchen windows open into a light shaft or closed yard where the smell may linger for hours

Table 41   *Recommended air exhaustion data*

Type of space	Volume required per person ft^3	m^3	Fresh air required per person ft^3/min	m^3/h	Air changes per person per hour
Aircraft	70–125	2–3·5	30	50	15–25
Cinema	200–300	6–9	30	50	6–9
Cold storage space					2–5
Conference room	250–300	7–9	40	70	8–10
Department store	300–400	9–11	15	25	2
basement	200–250	6–7	25	45	6–8
General office	500–700	14–20	15	25	1–3
Hospital:					
operating theatre	>1000	>30	>40	>70	2–4
general ward	300–500	9–14	25	45	3–4
private ward	>750	>20	25	45	2
Private office or home	>1000	>30	15	25	1
Public toilets					12–15
Restaurant	300–400	9–11	25	45	4–5
School	200–250	6–7	20	35	5–6
Theatre	200–300	6–9	30	50	6–9

and days. The tenants of other flats in the house will soon complain to the Local Authority. It is not easy to remedy the situation. An air conditioning unit working in reverse, or a duct leading up to and above roof level, will take away the smell, but these are expensive installations.

Natural draught ventilation is inadequate in large places and where people gather or work. Certain minimum numbers of air changes per hour are recommended for the maintenance of a healthy atmosphere.

An excellent help to ventilation is offered in Chapter 81 of the Municipal Code of Chicago relating to Ventilation. The Table of Ventilating Requirements comprises nearly 200 different locations and applications of domestic, commercial, and industrial ventilating requirements with exact specifications for the size of ventilating apertures, and minimum air supply or exhaust in ft^3/m per square foot of floor area. The pamphlet is available from the Municipal Bureau of Heating, Ventilation and Industrial Sanitation, Chicago.

Table 42

Specific gravity of some osmogenic gases and vapours

Air	1·000
Acetylene	0·907
Ammonia	0·596
n-Butane	2·085
i-Butane	2·067
Chlorine	2·486
Dimethylamine	1·521
Ethane	1·049
Ethylene	0·975
Hydrogen selenide	2·839
Hydrogen sulphide	1·190
Methane	0·554
Methylamine	1·080
Methyl chloride	1·785
Nitric oxide	1·037
Nitrous oxide	1·529
Ozone	1·658
Propane	1·562
Sulphur dioxide	2·264
Trimethylamine	1·996

If the air carries particulate matter, extraction of air by natural draught is wholly inadequate. It is ineffective also in all those cases where gaseous fumes of a density above unity pollute the air. Table 37 shows that with the exception of but a few, most osmogenic vapours are heavier than air.

Exhaust inlets should be near the ceiling, and fresh air inlets near the bottom on opposite walls or at least 7 m (20 ft) apart if the pollutant is lighter than air. The arrangement of in- and outlets is reversed for pollutants heavier than air. Where hoods are used at the intake ends of ducts these should be as close as possible to the source of fumes, because the velocity of air being moved decreases with the square of the distance between source and hood.

Of value are also the British Standard Institution's Code of Practice, CP 352, *Ventilation and Air-Conditioning*, and BS 2582: *Refrigerated Air-Conditioning Units*.

36 Electrostatic precipitation

This method is effective only on particulates, not with matter in the molecular state. Electrostatic precipitation will not remove osmogenic, or any other, gases from the air, but is useful where offensive effluvia are emitted by airborne dust or droplets.

Modern high-voltage machines can effectively deal with colloidal particulates, i.e., dust or droplets down to 0.005 μm size, or even less.

Using the Van de Graaff principle of linear acceleration, extremely small, even portable, d.c. generators in the range of 1 MV or more have become available commercially. It is sound practice to apply the high voltage in several steps so that the coarse particles are removed first, then the finer grades, and finally at maximum output, the finest dusts and droplets are precipitated.

The field strength or intensity E of the electric field between the two sets of wire or mesh electrodes in a precipitator is given by

$$E = -(dV/ds)$$

where V is the potential difference between electrodes separated by distance s. The force F acting on a charge q in the electric field, i.e., acting on a particle carrying charge q and travelling in the electrostatic field, is

$$F = qE \qquad \text{or} \qquad F = -q(dV/ds)$$

Factors q and s are fixed data specified by the designer of the electrostatic precipitator. They are invariable for any given machine. Remains the potential difference as the only variable. If the voltage is in the region of 5 MV the force F is sufficient to have a marked effect on large molecules. At still higher voltages, smaller molecules may become affected. Although the physics of molecular precipitation has, of course, been fully known for many years, the economics of these giant machines are preventing their application in such relatively paltry things as precipitation of osmogenic molecules.

Electrostatic precipitation, like several other techniques of extracting pollutants from the air, is but a 'collecting' technique offering only temporary relief, and producing a powerful secondary source of smell which must be disposed of by a more effective or more fundamental, process.

37 Condensation

It should be realized at once that condensation of offensive fumes is no solution at all, but only a more economic approach to dealing with osmogenic air pollution. Rather than handle large volumes of air, often hot air, containing offensive matter in a highly diluted state, the engineer will favour equipment for concentrating the pollutant by condensing it. The principle of condensation remains the same whatever the design of the condenser. There is the hot phase which carries the pollutant, and is bodily separated from the cold phase or the coolant. The hot phase always contains air, and in nearly all industrial applications, also carries steam. If the pollutant is water-soluble it will condense in the water. Precipitation concentrates solid particulate matter, condensation concentrates airborne liquid matter. Condensable vapours will be transformed into liquids which may, or may not, go into aqueous solution. The condensed waste matter, including fats and grease, will be collected in open lagoons or closed tanks and may—depending on the kind of osmogenic pollutant—be a lesser nuisance in the cold (condensed) state than when hot, or it may retain its offensiveness.

In any case, the discharging of the condensate will produce another smell hazard which, this time, comes from tanks and lagoons or, after discharge into sewers, from those. The problem of air pollution has been turned into one of water pollution. The same regulations apply to discharge from condensers as have been mentioned in the section on wet-scrubbing. In UK the Rivers (Prevention of Pollution) Acts apply, and the Ministry of Housing and Local Government leaflet 39/61, and subsequent issues, should be consulted for guidance. It is explicitly stated that the Common Law relating to nuisance is not affected by these Acts which refer only to the sanitary aspects of rivers, and to riparian rights (rights on river banks). When discharge into sewers is contemplated the Public Health (Drainage of Premises) Act, 1957, applies. Briefly, waste water must not be at a temperature above 110°F (about 44°C), and suspended solids should be within the range 200–500 ppm. Acidity below pH 6 is unacceptable, but alkaline waste up to pH 12 may be discharged directly.

Condensing, like wet-scrubbing, is a process requiring ample water and a well-functioning sewerage system.

Table 43   *Requirements of cooling water*

| Temperature at condenser outlet °C | Water required at 15°C | | | |
| | per lb (kg) of vapour | | per 1000 lb (1000 kg) of vapour per hour | |
	lb	kg	gallon	litre
15	230	230	23 000	230 000
20	75	75	7500	75 000
25	45	45	4500	45 000
30	32	32	3200	32 000
40	21	21	2100	21 000
50	16	16	1600	16 000
60	11	11	1100	11 000

The point made before that condensation does not necessarily reduce the offensiveness of a vapour by reducing its temperature, is borne out in the next Table which is compiled from data by RONALD (146). The quantities quoted here are relative values, referring to vapour at 100°C, and to vapour at 15°C. The values refer to vapours of trimethylamine $(CH_3)_3N$, and to hydrogen sulphide, H_2S, from fresh fish and from stale fish when both have been boiled at the higher temperature.

Table 44   *Effect of cooling by condensing from 100°C to 15°C on the yield of offensive odours from fish*

| | Trimethylamine from | | Hydrogen sulphide from | |
| | fresh | stale | fresh | stale |
		Fish		Fish
Reduction %	22·75	10·2	21·4	22·4

The condensation (dew) point is that temperature at which a vapour is transformed into a liquid. For a gas to be condensable its condensation point must clearly be above the temperature of the cooling medium available. The condensation point lies below the boiling point on the temperature scale; the heat transfer in the process of condensation has been discussed by OTHMER (157).

The efficiency of the condenser varies with its type, with the kind of material that is the source of smell, and with the type of vapour itself. In processing meat, for hydrogen sulphide the spray condenser is superior to a surface condenser, but for di- and trimethylamine, both are about equally effective. In processing fish, for dimethylamine a surface condenser is preferred, but for the other gases, a spray condenser should be used (146).

A hot effluent has a considerable oxygen requirement. Cooling it down to below 45°C produces an effluent better than to BOD standards. Volatile organic matter, such as the above smells, has a very offensive odour which becomes worse in the presence of ammonia or amines. All are water-insoluble. Hydrogen sulphide and ammonia combine to form ammonium sulphide which has an offensive smell of its own.

$$2NH_3 + H_2S \rightarrow (NH_4)_2S$$

38 Masking

The masking of one odour by another is as old as man's notion of offensive smells. And just as little as it helped thousands of years ago will it relieve unpleasantness today. This view was strongly confirmed at a Symposium on Odour held at Doncaster Technical College on 14th March, 1968: 'Adding a "stink" to hide a "stink" profits only the salesman of the material' (229).

To appreciate the full truth of this statement some basic facts will be repeated.

(1) An odour is perceived only after osmogenic particles or molecules have entered the nasal cavity and passages, and have reached the olfactory area near the root of the nose.

(2) In masking, another osmogen of specified characteristics is released into the polluted air. A person inhaling this mixture may, for a time, experience a sensation of 'no odour', yet this is not caused by making the offending stuff nonosmogenic (as by oxidizing it), but by causing the person to inspire two—instead of the original one—smells simultaneously. But it is characteristic of the masking substance that its odour is predominant with regard to the offending odour.

(3) There are two possibilities of interaction. Either the predominance of the masking agent is such that it overcomes (masks) the offensive smell by its own perfume which may be more or less innocuous to people; or it partially paralyzes the olfactory nervous system to the perception of the offensive odour, an effect which is commonly spoken of as neutralizing the odour. The psychological result of 'no odour' is of questionable value since the effect is attained under unstable conditions, and because the offensive odour persists in spite of the treatment. In fact, it is the observer who is 'treated', not the odour.

Whether such a method is justifiably applicable except, perhaps, in special circumstances, must be decided by the expert.

(4) There is another point to watch, namely the point of application of the masking agent. As in true masking, one smell is pitched against the other, a satisfactory result can only be obtained if the quantitative as well as the qualitative ratios between the two are correct and constant. It is not too difficult to establish some sort of stable conditions between the two compounds if the polluted air is confined to a room, but it is impossible in unrestricted space, e.g., in the open, where winds will carry the offending smell (usually in the molecular state) and the masking agent (in liquid particulate form of various droplet sizes) at different velocities and along different trajectories so that no observer will be able to inhale the correct mixture of smell and agent anywhere.

The masking of odour gives rise to some concern about the health of people who inhale the various chemicals used in these agents. Although the manufacturers doubtless take every care to ensure their preparations are harmless, no one can give an assurance that there might not be long-term effects. Modern man lives in an environment of which sprays and scents and deodorizers, and so on, form an integral part. When a fly moves, a jet of spray is squirted at it. The bathroom is steeped in a scent of 'natural freshness', and the perfume of bath cubes, so the print on the packing promises, will make the Queen of Sheba green with envy. Most of these marvels of personal chemistry are dispensed as an aerosol from a spray can, and with every jet of active ingredient is also released a goodly portion of propellant—and man inhales the lot. A study of US Pat. No. 3 328 312, granted 27th June, 1967, reveals the composition of one active ingredient and its propellant. The former is perchloryl fluoride, the latter a mixture of dichlorodifluoromethane CCl_2F_2 and trichlorofluoromethane CCl_3F which is Freon or Frigen. A mixture, said to be effective in diluting repulsive odours so that they were no longer noticeable, contained 0.4% by weight of perchloryl fluoride.

A perchloryl is a chloryl (monovalent ClO^- radical) which contains more chlorine than the corresponding ordinary chloryl. The reagent as well as the propellant are chlorine-fluorine compounds. Freon is a nearly odourless refrigerant only used in closed systems—except when used as a propellant—which explains why its effects on man have not been extensively studied with a view to long-term action. For perchloryl fluoride, the maximum acceptable concentration (MAC) is 3 ppm which is 13.5 mg/m^3. A typical pressurized can contains about one-half gram of perchloryl fluoride. If the whole bottle should be released at once into a room of average size, say 50 m^3, the concentration of perchloryl fluoride in the air of this room would be $0.5/50 = 10$ mg/m^3. This is very close to the acceptable limit concentration (13.5 mg/m^3) relating to immediate effects on man. How the contact of lung tissue with chlorine and fluorine compounds in nearly maximum concentration would work out over long periods of time is not known.

In many instances, additives are blends of aromatic oils. Depending on temperature, pH, wind direction and velocity, the effect of an additive will greatly vary. On hot days or in the presence of strong acids, even in traces, masking will not be a success, because the reagents decompose quickly under these conditions.

Additives may be mixed with the raw material before processing at a rate recommended by the manufacturers, but generally between 2–4 oz/t (= 50–100 g/t). This has the disadvantage of tainting the finished product more or less; or the agent is added to the scrubbing water at about 2–4 oz/gall (= 10–20 g/l). It may also be added to liquid effluents (sewage), at about 1–3 ppm by weight, or to gaseous effluents at the rate of 10 ppm–20 000 ppm.

Lagoons and tanks, it is suggested, may be prevented from emitting offensive odours by so compounding a floating chemical screen that any emissions from the liquid surface must pass through the masking agent. The rate of application is about 5 ppm.

Aerosols are dispensed either by hand from pressurized plastic containers, or by small pumps mounted on poles producing continuous jets of suitable shapes. Masking or any procedure using an additive must not be employed when the osmogen also is pathogenic or toxic, or carries pathogenic microbes.

Chlorophyll $C_{55}H_{72}O_5N_4Mg$, the green colouring matter of vegetation, has been credited with powers of deodorization. The principal thought of this hypothesis was based on the fact that chlorophyll assists in oxidizing obnoxious matter. Chlorophyll interacts with carbon dioxide and water by combining these to sugar whereby oxygen is released

$$\text{chlorophyll} + 6CO_2 + 6H_2O + 674 \text{ kcal} \rightarrow C_6H_{12}O_6 + 6O_2$$

This is the basic process of photosynthesis.

Chlorophyll has been widely advertised—and used in households—for destroying unpleasant smells. WEAVER (338) of the National Bureau of Standards has proved that using a wick dipped into an aqueous or alcoholic solution of chlorophyll cannot succeed in deodorizing the air. Other authors have supported this claim by showing that chlorophyll has no effect on osmogenic molecules.

A wick moistened by chlorophyll will absorb—HAINER (339) prefers to say adsorbs—a proportion of hydrogen sulphide molecules which may be in the air surrounding the wick. The stress here is on the definition 'surrounding' as there is no distance effect. Other H_2S molecules will migrate towards the immediate vicinity of the wick since this has become depleted of H_2S molecules. The psychological effect of this wandering off of the molecules in the direction of the bottle containing the wick is interpreted by the mind as a reduction of smell intensity.

This preference of a chlorophyllated wick for H_2S molecules has been demonstrated convincingly by work in the National Chemical Laboratory, Teddington (340). A wick soaked in chlorophyll, when placed near silver, copper, and their alloys, in an atmosphere containing hydrogen sulphide vapours, will prevent the metal from becoming tarnished. A water-soluble chlorophyll preparation will take up half its own weight of H_2S in 15 hours; but the action becomes apparent only in its immediate vicinity.

In recent years, chlorophyll has been replaced by, or used in combination with, other substances all reputed to have deodorizing properties. Careful consideration should be given to any of these claims before their use is contemplated.

39 Combustion

Combustion is the rapid chemical combination of oxygen with combustible matter (fuel). When injected into a combustion chamber such matter (solid, liquid, gaseous) absorbs heat and evolves volatile hydrocarbons which ignite and burn off more or less completely. Temperature, proper mixing of the ignitable vapours with the oxygen from the air, and a specified length of time must be available if combustion is to be complete. In which case for most normal combustibles the only products will be CO_2 and H_2O.

Table 45 *Combustion properties*

Fuel	Combustion process	Approx. ignition temperature °C	m^3 of air per m^3 of fuel	Combustion end products (waste)
C	$C + O_2 \rightarrow CO_2$	400	43	CO_2
H_2	$2H_2 + O_2 \rightarrow 2H_2O$	590	1	H_2O
CH_4	$CH_4 + 2O_2 \rightarrow CO_2 + 2H_2O$	670	3	$CO_2 + H_2O$
C_2H_4	$C_2H_4 + 3O_2 \rightarrow 2CO_2 + 2H_2O$	600	4	$CO_2 + H_2O$
S	$S + O_2 \rightarrow SO_2$	250	16	SO_2

Combustion is sometimes suggested as an efficient means of 'burning the smell'. This is perhaps an attractive proposition to the owner of both a furnace and an objectionable smell, but to the neighbours waiting for the relieving conflagration of offensive matter in a holocaust it turns out to be a damp squib.

The reason for the disappointment is basic physics. It would be uneconomical to run a furnace designed for doing a particular job particularly well, on a higher temperature than is required for doing the job; it would mean a greater fuel consumption and shorter furnace life. The amount of air (which contains only one-fifth of its volume oxygen) required for complete combustion to carbon dioxide and water is not really very big, and it can be preheated. Suddenly, a considerable volume of cold polluted air passes through the combustion chamber where the pollutants are expected to become oxidized. This reduces the temperature and frustrates combustion. Waste products from incomplete combustion are mostly evil smelling compounds, making the situation worse than it was.

Further, any oxidation taking place demands the initial breaking down of atmospheric molecular oxygen O_2 into two nascent oxygen atoms. It can easily be shown that even in a furnace operating at 1200°C (involving special structural protective methods) the energy falls well short of that needed for this purpose: namely for each gram-molecule of oxygen, 487 kilojoules.

The polluted air is heated relatively slowly on its way to the combustion chamber. This means that the pollutants undergo dry distillation, and some of these products may have ignition temperatures far above those attained in the furnace. These substances will not be oxidized at all and pass into the open unchecked.

Hydrogen sulphide will split into water and sulphur, $H_2S + O \rightarrow H_2O + S$, upon passing through the furnace. The sulphur may either be deposited in the open as a whitish-yellow powder or oxidize to SO_2 creating a new hazard. The sulphur dioxide may combine with water vapour to form sulphurous acid (a strong corrosive), or it may remain in the air as it is. This is a highly undesirable choice, and not made any more attractive by saying that unless the waste effluent also contains a minimum of 15 % carbon dioxide, SO_2 will certainly form in the air. There is no comfort from this stipulation, because 15 % CO_2 will give ample cause for the formation of black smoke which in UK is a violation under the Clean Air Act.

Combustion can take other forms, for instance, that of catalytic combustion. This is a change in the conditions of the process. In the presence of the catalyst combustion will take place at a temperature lower than without a catalyst. Catalytic combustion is flameless in most cases, and capable of burning the pollutant at various air concentrations. A large surface-to-volume ratio is an essential quality of a catalyst regardless of its specificity. Catalysts are very sensitive to foreign matter for which the catalyst is not specific and which cannot undergo catalytical combustion. These substances are known as 'poisons' and ruin a catalyst. Modern catalysts in the form of beads have high surface-to-volume ratios, often as much as $750 \text{ m}^2/\text{g}$ (5 acres per ounce) of substance (270). Catalytic ceramic rods have less effective ratios, perhaps $200 \text{ m}^2/\text{g}$. Suitable catalysts for combustive oxidation, are:

Pollutant	Catalyst	Combustion product
Amines	Silver	Nitrites
Aldehydes	Vanadium	Maleic anhydride
Mercaptans	activated	Sulphides
	Aluminium oxide	

for instance: CH_3NH_2 : Ag → CH_3NO_2
              methylamine   silver   methyl nitrite

Only oxidizable gases can be burnt either straight or by catalytic action. Non-combustibles should first be removed by adsorption on activated carbon or other suitable adsorbents, or they will pass through the furnace unchanged.

In catalytic combustion, contact and contact time are the determining factors. As the catalyst has no distance effect each molecule of combustible matter must make bodily contact with the surface of the catalyst. The configuration of the catalytic elements, and their arrangement in a pattern across the entire section of the gas flow will determine the efficiency and effectivity of the display. Eddy currents should be avoided as they give molecules an opportunity to flow between, instead of in contact with, the elements of the catalyst, thus missing the contact. The arrangement of elements in the direction of flow provides control of the required contact time. Catalytic combustion is an exothermic process. It can occur under conditions which would not maintain ordinary (flame) combustion. When only traces of combustible pollutants are in the air passing through the furnace there is a possibility that ignition temperature (however low it may be, e.g., only 200–250°C) may not be reached. Heat must then be provided by an auxiliary source (heat exchanger, flame, electrically heated wire) which is automatically switched off when catalytic combustion is operating. Thermal switches (cut-outs) of a type suitable to the purpose are then fitted just below the catalyst.

Catalytic combustion is at its best and most economic, when sufficient combustible matter is present to make the process a steady source of heat. Sudden changes in the supply of odoriferous combustible waste gases which cause surges and lapses in the volume and/or quality of burnable gases are not conducive to economic working. Such may be the case with oil refineries, amine production plant, or any other chemical works where peaks of concentration or quantities are released at unpredictable or long intervals. Concentrations of combustible pollutants should be above 1000 ppm. Below that, catalytic combustion becomes uneconomical.

The theoretical rise in temperature of polluted air owing to catalytic combustion is

$$\Delta T \approx 29 \times R$$

at 20°C ambient temperature, where R is the reading taken with a combustible-gas indicator calibrated in per cent ot the lower explosive limit. The limits of flammability indicate the range of combustible-in-air concentrations within which direct combustion will take place. Below or above these limits combustion will occur only if excessive heat, or a catalyst, is applied. Care should be taken to make sure that mixtures of compounds with air, oxygen, or nitrogen dioxide, do not form explosives although each compound may be nonexplosive by itself. Catalytic combustion of dimethylamine (synthetic fibres, fish meal) has been successfully used on an industrial scale (271, 272).

Both catalytic combustion and thermal afterburning have been used in a multiplicity of forms to reduce the problem of exhaust from motor vehicles. Considering that the average car discharges some 6–7% of its fuel unburnt, the magnitude of the problem is really shattering. Several devices have been constructed and run in cars for a small number of years, but they have proved in many a case to be really helpful. Afterburning takes care of both incompletely burnt exhaust gases from the cylinders, and unburnt gases and oil vapours from the crankcase (60).

On a more concentrated, yet by no means lesser scale, ranges the combustion of domestic refuse in incinerators. Rural and smaller Local Authorities will not always go to the trouble and expense of purchasing, operating, and maintaining a proper incinerator, but burn the collected refuse and rubbish on the waste heap. The result does not always meet with the approval of the community or that part of it which suffers from the smell, dust, and smoke of the conflagration. Even the use of incinerators does not always present the desired solution, owing to the blanketing of fire and uncontrolled admission of air caused by feeding the furnace in batches and at irregular intervals. A continuous feeding process is suggested (274) as it will keep the incinerator temperature and flue gas rate more uniform. The more constant flow of waste gases will improve

Table 46   *Flammable concentrations of gases and vapours in air*

	Lower limit %	Upper limit %
Acetylene	3·0	73·0
Alcohol	4·0	14·0
Benzene	2·7	6·3
Blast furnace gas	36·0	65·0
Carbon monoxide	15·0	73·4
Chlorine dioxide	10·0	—
Diethylamine	1·2	8·0
Diethylether	2·9	7·5
Dimethylamine	2·8	14·4
Ethane	2·5	5·0
Ethyl ether	1·1	—
Hydrogen	10·0	66·0
Hydrogen sulphide	4·5	45·5
Methane	5·5	14·5
Monoethylamine	1·8	10·1
Monomethylamine	4·95	20·75
Natural gas	5·0	12·0
Ozone in oxygen	10·0	—
Petrol vapour	1·5	6·0

the design of fly-ash collectors and rationalize their sizing, as well as the installation of odour control equipment at a reasonable cost.

If sewage sludge cannot be disposed of by mixing it with domestic refuse (dust, vegetable garbage) and turning it into fertilizer for agricultural use, there is little one can do with it, except burn it. The pyrolysis of any waste is a dangerous task as explained above, but the burning of sewage slude is particularly dangerous, because of its sulphur content. Unless furnaces are used which have been specially designed for the purpose, air pollution on a remarkable scale will be caused. Though smoke is inevitable unless electrostatic precipitation is used, it will be osmogenic pollution through partly oxidized compounds which gives serious offence. McAteer has designed a sewage sludge incinerator featuring a closed burning grate. The designer claims freedom from offensive smells by complete combustion. Heavy hydrocarbons are eliminated (275). These are oxidized in a space just above the fuel bed in the presence of an additional (secondary) air supply to provide sufficient oxygen for the purpose.

40 Ultra-violet irradiation

Deodorization by ultra-violet irradiation of the source of odour is the only method which deserves rightly the descriptive name 'deodorization by destruction'. The term 'destruction' must be understood in the organoleptic sense, i.e., in relation to the mental interpretation of physiological reactions; for the word itself has no precise meaning either chemically, or physically. The basis of organoleptic decision is the presence or absence, of stimulation of the sense organ. It is quite irrelevant whether a substance is of such molecular structure that the particular sense organ cannot react to its presence, or whether a substance which would stimulate the sense organ into activity, has so been changed in molecular structure that it can no longer do so.

Man's olfactory perception does not cover oxidized substances, hence this characteristic has been made the principle of destructive deodorization. By tagging an oxygen atom on to molecules which produce an aversive olfactory response they lose the faculty of exciting man's sense of smell. The oxidized molecules cannot be detected organoleptically, hence they do not bother the observer.

Oxygen atoms (nascent oxygen) are generated from environmental air by irradiating it with energy emitted from low-pressure mercury vapour discharge tubes made from quartz. Ordinary, or commercial, quartz (not quartz glass) is quite transparent to the short wavelengths which split the oxygen molecule. These tubes emit more than 96 % of their total ultra-violet output at 2537 Å, but this wavelength has very little effect in causing the oxygen molecule to dissociate into two nascent oxygen atoms. This is done by wavelengths shorter than 2000 Å, preferably by radiation of 1849 Å, which is also emitted.

Radiation shorter than 1849 Å is generated in the discharge column within the tube, but cannot penetrate the quartz envelope, which has a lower limit of optical transmissivity at just below 1849 Å hence the low

Table 47   *Typical spectral emission from a*
low-pressure mercury-vapour-in-quartz generator

	Wavelength Å	Intensity $\mu W/cm^2/1m$
Ultra-violet	1849	2·47
	2482	0·03
	2537	120
	2652	0·07
	2804	0·02
	2894	0·05
	2967	0·22
	3024	0·07
	3132	0·71
	3342	0·04
	3650	0·65
Luminous	4047 violet	0·47
	4358 blue	1·20
	5461 green	1·05

output at 1849 Å (2·47 $\mu W/cm^2/1m$) as against the emission at 2537 Å (120 $\mu W/cm^2/1m$) which spectral line is well transmitted through quartz. The envelope is well suited to produce simultaneously and inseparably 1849 Å and 2537 Å emission and, consequently, is a source of microbicidal rays, nascent oxygen, and ozone. By doping commercial grade quartz with ions of iron, or with titanium dioxide, the lower limit of transmissivity is raised in proportion with the quantitative doping, and can be raised to a level beyond 2000 Å, when ozone generation will be inhibited. The same end is attained when envelopes of hard (silica) glass, often known as quartz glass, are used, but the transmissivity in the microbicidal region is also reduced.

Output of ultraviolet energy is dependent on environmental temperature, but more so if a cold cathode tube is used. It produces its maximum output at 2537 Å with temperatures ranging between 10 and 20°C. At lower and higher temperatures the output drops. The 1849 Å output increases with increasing temperature, a practical limit being 50–60°C. At low temperatures the 2537 Å line outweighs the output at 1849 Å.

At about 27–28°C both outputs are quantitatively the same, and at still higher temperatures the ozonogenic radiation leads the microbicidal emission. Optimal working temperature is 27–28°C. With hot cathodes these ranges are less pronounced as also is the effect of environmental temperature.

The rated life and efficiency of low-pressure tubes, after allowing for the initial burning-in period of 50–100 hours, is 5000 hours, although cold cathode generators are known to have worked for 10 000 hours continuously without failing once. However, the same can be said of the hot cathode tube if it has a reliable electrode system. Hot cathodes, by the sheer working principle involved, lose efficiency slightly faster than cold cathode tubes. Efficiency is reduced to 75% after 6000 hours for cold cathode tubes, but after 5500 hours for hot cathode generators. And after 7500 hours the hot cathode burner operates at an efficiency of 64%.

The end of the useful life period is not in consequence of electrical breakdown which is a defect in the discharge tube, but because of optical breakdown occurring in the envelope. A quartz tube which has not yet been taken into operation exhibits a fairly strong bluish fluorescence under impact of filtered 3650–63 Å radiation. As operational periods add up, the ability to fluoresce diminishes progressively. The formerly crystal clear quartz takes on a purplish-mauve tint, an effect of solarization, which deepens as the useful life progresses. Transmissivity of the solarized quartz diminishes for all spectral ranges, but especially for those of short wavelengths. This means that the envelope becomes useless in the range of oxygen splitting rays, but may still serve as a generator of microbicidal radiation. When the shade of the quartz turns a deep purple the tube should be discarded.

Loss in fluorescence and the incidence, and subsequent change in hue and intensity of discoloration are the outward signs of molecular transpositions in the quartz envelope. They are due to changes in the crystalline structure caused by rays shorter in wavelength than the lower limit of transmissivity. Because of this darkening, the shorter wavelengths cannot be transmitted from the interior of the quartz tube, where they are generated, to the environment of the tube outside the envelope. These rays are absorbed by the quartz. The energy E of radiation is inversely proportional to wavelength λ and is formulated as

$$E = hc/\lambda$$

where c is the velocity of light. The shorter the wavelength, the greater the amount of energy absorbed by the quartz, and the greater the displacement of molecules in the crystalline structure.

Electrical failure of an ultra-violet generator may occur at any moment of its life, but is rare in reliable products. The discharge within the tube occurs in an atmosphere of mercury vapour and a mixture of neon and argon. The main purpose of the gases is to form a conducting bridge across the electrode gap at the moment of starting from cold, and during the initial burning-in period of, usually, 3–4 seconds. This is the time required for formation of hot mercury vapour, and before it contains sufficient molecules to produce a metallic bridge across the gap instead of the gaseous one. The number of gas molecules which allow the tube to be started from cold, collide during operation with molecules emitted from the metal cathodes by a process known as sputtering. This collision reduces the number of gas molecules and striking the tube from cold will get progressively harder. Within a short time the tube must be replaced. Sputtering is caused by a variety of reasons and produces as visible evidence of its existence a black mirror-like coating on the inner wall of the tube around the electrodes (electrode chambers). The reduction in molecules takes place by the metal molecules hitting gas molecules and causing them to move with them as they travel away from their source towards the cooler wall of the tube. This phenomenon is called the hardening of the tube and should not, normally, occur until during the latter period of the tube's life.

Owing to losses occurring in the optical and in the electrical domains of ultra-violet generation, each watt of ultra-violet output requires about 15 W of electrical input for a cold cathode tube. This is a 6–7 % efficiency.

Ultra-violet generators can be made to give any required output, but geometrical considerations of installation do strongly suggest that a number of small output generators distributed effectively over the area or through the volume where destructive deodorization is to take place, is preferable to using one, or a few, high-power generators. If these were fitted at either one point (which results in insufficient exposure time) or well spaced (which produces too great distances between generators) radiation would not be powerful enough to oxidize the osmogenic molecules.

The shape of a low-pressure discharge tube is essential. When produced as a long straight tube its intensity diminishes inversely proportional to the first power of the distance between tube and irradiated area; when ring-shaped it can be regarded roughly as a point source, and intensity varies inversely with the square of the distance. The linear shape of ultra-violet generators is preferred, the annular form being more expensive to produce and giving unsatisfactory distribution of radiation. For practical calculations, the intensity may be considered following a first power law although there are slight deviations as ANDERSON (264) has pointed out.

Table 48 *Intensity and distance of straight line sources*

| Distance | Relative intensity | |
cm	measured	calculated
15	1·000	1·000
25	0·670	0·670
30	0·480	0·500
40	0·375	0·400
50	0·325	0·333
60	0·225	0·250
75	0·175	0·220
90	0·148	0·172
100	0·125	0·150
125	0·092	0·121
150	0·074	0·100
175	0·063	0·086

The reason for this deviation from the rule is the reduced intensity towards the ends of the tube, i.e., towards the cathodes. The farther away from the tube measurements are made, the more pronounced becomes the effect.

Ultra-violet rays are emitted from the tube in a radial direction. This distributed pattern is not suited to the application, and is modified by using reflectors. The surface of the reflector is of much greater importance than its shape which may be flat or curved (268). The optical characteristics of the reflector surface depend on the wavelength of radiation. This is taken to be in the 1800–2600 Å range. There are several suitable materials available for the construction of reflectors, but for industrial applications maximum reflexion in the specified region, and a reasonably long useful life, is offered by electrolytically brightened super-purity aluminium. In strongly corrosive atmospheres and after long periods of use, the aluminium surface will become oxidized, forming a fine ceramic film of a whitish colour which will protect the metal against further deterioration, but will greatly reduce reflectivity. The ageing of the aluminium reflector is wavelength dependent.

Age	Percentage reflectivity at		
	1500 Å	1600 Å	2200 Å
New	71	78	91
10 000 h	51	64	90
Decrease %	28	18	1

Materials other than aluminium are unsuitable as reflectors in the 1800–2600 Å region unless reflectivity coefficients of below 25 % are considered satisfactory for the purpose. Electrolytically brightened (brytalized) super-purity aluminium shows hardly any reduction in reflectivity after 9 months' exposure to ultra-violet radiation.

The geometry of irradiation is simple and mostly follows the general law obtaining in illumination (Figures 41 and 42). Only a small number of configurations need be discussed as the designer will meet these same conditions most frequently in practice.

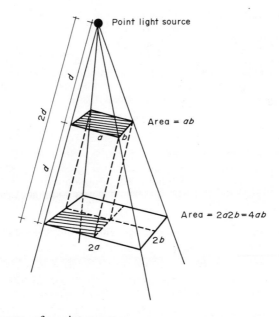

Figure 41   The geometry of a point source.
At twice the distance of a plane from the source the illuminated area grows by the square of the distance (d^2) and the intensity of illumination is proportional to $(1/d)^2$. The inverse square law obtains.

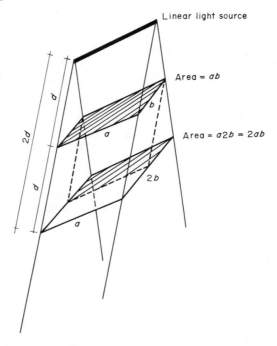

Figure 42   The geometry of a linear source.
At twice the distance of a plane from the source of illumination the illuminated area grows linearly with distance, and the intensity of illumination is linearly proportional to $(1/d)$. A first power law obtains.

Vertical wall: The ultra-violet sources should be mounted at the middle of the wall, not at the top. If more than one row of generators is required, an arrangement as shown in Figure 43 is preferred.

Ducts and irradiation chambers: The tendency to use ultra-violet sources in deep reflectors should not be continued. The entire cross-section of a chamber, or duct, must be filled with radiation, to prevent blanks through which osmogens could escape without passing through the radiation; so no reflectors should be used, Figure 44(a). Restriction in the spread of the radiation, due to shaped reflectors, is shown in Figure 44(b).

Either for technical or economical reasons, it may not always be possible to separate molecular from particulate pollutants before the air is admitted to the irradiation chamber. If this is the case, the quartz tubes may become coated with the solids or liquids carried along and, even after a short time of operation, lose their effect due to the heavy coating of particulate matter on the surface of the tubes. It may be tar droplets,

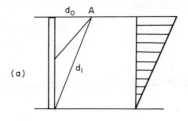

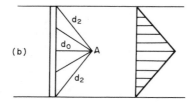

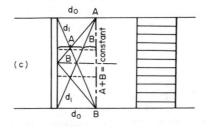

Figure 43   Irradiation of a vertical wall by a linear source (perpendicular to the plane of the paper).
(a) The source A is arranged at one end of the wall. Intensity is highest right under the source (d_0) and lowest at maximum distance (d_1) from the source.
(b) The source A is arranged at the middle of the wall. Intensity is highest right under the source (d_0) and lowest at maximum distance (d_2) from the source, but $d_2 < d_1$, therefore the intensity of illumination is better than in (a).
(c) Two sources A and B are arranged at either end of the wall. The intensity of illumination at any point along the wall between A and B is constant and equal to the sum of intensities A and B from both sources obtaining at any particular point.

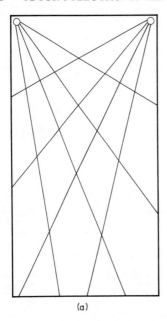

(a)   (b)

Figure 44   Two identical rooms are illuminated with identical sources except for the fact that in (a) the rays are not controlled in their distribution, whereas in (b) the sources are situated in reflectors which restrict the spread of radiation. The shaded portions in room (b) are insufficiently protected by radiation.

hot grease or fat droplets, dust, moisture, or—mostly—a mixture of several. Although it is technically unwise to allow such a state to exist or, even, continue it is found only too often in practice. To prevent tubes and reflectors from becoming coated with airborne dirt these elements are best recessed 5–10 cm thus withdrawing them from the direct impact of the dirt. A jet of compressed air should be provided at the deepest point of each recess blowing in a direction away from the tube towards the centre of the duct or chamber. This will prevent eddy air currents around and in the recesses causing a deposition of dirt on the generators, Figure 45.

The dimensions of an irradiation chamber are calculated from the total air volume moving through the system per unit of time, and from the cross-section of the duct leading into the chamber.

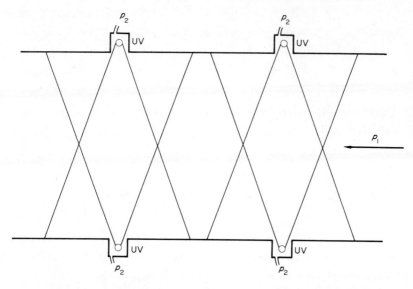

Figure 45   Arrangement of sources of ultra-violet radiation UV in recesses across a duct or irradiation chamber, if particulate matter as well as offensive fumes are carried. The polluted air at pressure p_1 moves in the direction of the arrow. Compressed air at pressure $p_2 > p_1$ is introduced behind the ultra-violet tubes to prevent the particulates from settling on tubes and reflectors.

V = volume (m³/sec) of air to be irradiated,

q_d = cross-section (m²) of duct leading into chamber,

c_d = velocity (m/sec) of air in duct,

β = ratio of duct cross-section, q_d, to cross-section of chamber, q_c, where

$$\beta = q_d/q_c < 1$$

This is the factor by which c_d is reduced to c_c (air velocity in the chamber). $c_c = \beta c_d$.

l_c = length (m) of the irradiation chamber on the condition that t seconds, usually 4–6 sec, are allowed for the air to remain in the chamber. $l_c = c_c t$.

acc = air changes in the chamber; this is the ratio of the total volume, V, to the volume of the chamber which is $q_c l$, thus acc = $V/q_c l$.

(*Note:* the relationship of all values remains valid if the quantities are expressed in any other system, for instance, in ft, ft², ft³.)

Example: An air flow of $8 \text{ m}^3/\text{sec}$ through a duct of $0 \cdot 2 \text{ m}^2$ cross-section should be retained in the chamber for 5 sec.

Given:
$$V = 8 \text{ m}^3/\text{sec} (= 480 \text{ m}^3/\text{min}),$$
$$q_d = 0 \cdot 2 \text{ m}^2,$$
$$c_d = V/q_d = 8/0 \cdot 2 = 40 \text{ m/sec},$$
$$t = 5 \text{ sec},$$

Assumed: $q_c = 2 \text{ m}^2,$

Thus:
$$\beta = 0 \cdot 2/2 = 0 \cdot 1 \text{ (which is } < 1),$$
$$c_d \beta = 40 \times 0 \cdot 1 = 4 \text{ m/sec} = c_c,$$
$$l = tc_c = 5 \times 4 = 20 \text{ m},$$
$$q_c l = 2 \times 20 = 40 \text{ m}^3 \text{ (volume of chamber)},$$
$$\text{acc} = 480/40 = 12/\text{min}, \text{ i.e., 12 air changes per min in the}$$
chamber.

Inlet and outlet ports should be at opposite ends of the chamber whose design involves more than the making of a square box. The air passing through the chamber should be maintained in a turbulent eddying motion to make sure that each molecule of odour is carried close to the ultra-violet generators at one time or another. Suitably arranged baffles are quite satisfactory in attaining a controlled air flow as long as the designer makes sure that no baffle casts a shadow across the air stream, thus shielding the osmogens from the radiation. Without baffles the air will flow straight from inlet to exit and arrange itself into a pattern of a stagnant outer mantle of air hugging the walls and other contours of the chamber with an inner shaft of air rushing through the chamber at a speed much higher than calculated.

In order to maintain a fairly even distribution of ultra-violet intensity throughout the chamber, a limit of 2 m (7 ft) is tacitly assumed for its width. This is because all energy data of ultra-violet generators are given over a measuring distance of 1 m. As the tubes are mounted on the outside of the chamber walls, tubes on opposite walls are only exceptionally allowed to be mounted at more than 2 m distance so that the emission data apply in the axis plane of the chamber by simply adding up the emission of any two opposite tubes.

It is an advantage not to arrange the generators on opposite walls in exact mirror positions, but by offsetting them by one-half the distance between two immediately juxtaposed generators. This gives a better spread of radiation and prevents part of the energy emitted by one generator on one wall to be absorbed in the discharge column of another generator mounted exactly opposite on the other wall. A mercury vapour discharge not only emits radiation within a certain spectral section, but also absorbs the same range of radiation.

Maximum temperature of the air passing through the chamber should not exceed 42-45°C (about 110°F) to prevent the quartz tubes from overheating and causing the spectrum to shift towards longer waves, i.e., causing a loss of output.

The ventilating system cannot be upgraded by simply increasing both the pulling power of the fan and the volume of air throughput. A system of true destructive deodorization requires the polluted air to remain for a specified period of time in a predetermined location (the chamber) within the system, thus an increase in air speed and the necessity for having a certain residence time, are antagonistic requirements. These can be resolved only at excessive cost. A different approach must be found.

As residence time is a specification which allows of no modification it is only the overall volume of air flow which lends itself to adaptation. Air volumes are calculated on the basis of the number of changes per hour, acc, the value of the acc depending on the use of the building, the concentration of smell, and type of neighbourhood into which the (foul) air is discharged and from which clean air is injected into the building—if clean air is injected by a controlled process, and not by draught. Usually, it is only the foul air which is exhausted, and the outside 'fresh' air is left to its own devices to find a way into the building, by virtue of natural *horror vacui* (a term coined by Otto v. Guericke, physicist, 1602–1686).

When the polluted air in the building is reliably and irreversibly deodorized the value of acc can be significantly reduced, even down to $\frac{1}{4}$ of the originally specified figure. This opens the way to carrying out a programme of air deodorization on an economical basis, and to making it an amenity rather than a burden.

Rooms: Where ultra-violet generators are used in a room usually occupied by persons or animals, e.g., doctors' waiting rooms and surgeries, incurable wards in hospitals, animal rooms in hospitals, pathology laboratories, it is not only essential that the ozone output be kept below 0·1 ppm, possibly 0·05 ppm, but also that no ultra-violet rays, direct or reflected, can reach the eyes of persons or animals. Ultra-violet radiation below 3200 Å is absorbed by the mucous membranes lining the inner aspect of the eyelids (conjunctiva), and lead to a more or less severe, but very painful conjunctivitis (inflammation of the mucosae of the eye). The sensitivity of the biological structures increases with decreasing wavelength.

Table 49   *Relative conjunctivitis factor* (337)

Wavelength Å	Causative factor %
2399	100
2446	85
2482	75
2537	60
2576	50
2654	39
2675	37
2699	35
2753	30
2804	20
2894	10
2925	7·5
2967	5
3024	4·25
3132	3·75

Exposure of the eye to the 2537 Å radiation will cause irritation at the minimum level of 40 μWmin/cm^2, but the actual value is dependent on individual factors. The amount of blood in the mucosae is an important item because blood heavily absorbs this radiation, hence acts as a protective screen.

By multiplying the causative factor (percentage factor) of sensitivity, Table 49, with the spectral emission factors at corresponding wavelengths, Table 47, the values represent the weighted conjunctivitis factors of an ultra-violet generator described in Table 47.

It is obvious that in practice only the 2537 Å line is causing a health hazard. Considering the 40 μWmin/cm^2 level of energy striking the eye, an intensity of 7200 μWmin/cm^2 with the generator at 1 m distance from the eye need irradiate the mucosae of the eyelid only for about 0·33 sec (about the duration of a blink straight at the tube from 1 m distance) to produce an irritation. Protection of the eyes of man and animals in a room where ultra-violet rays are present, is essential.

Depending on the spread of radiation defined by the shape of the reflector, and height below ceiling, the ultra-violet generator will protect different sections of the room by radiation and nascent oxygen, Figure 46. Various aspects of reflector design are illustrated in Figure 47.

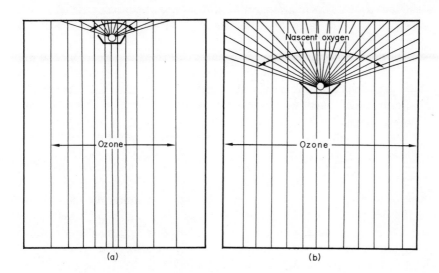

Figure 46   Effect of distance of a source from the ceiling on volume of irradiated air, and on amount of nascent oxygen and ozone produced.

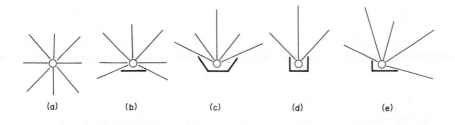

Figure 47   Effect of differently shaped reflectors on the distribution of radiation.
(a) No reflector, omnidirectional pattern of rays.
(b) Flat reflector.
(c) Flat trough reflector.
(d) Deep trough (box) reflector.
(e) Asymmetrically shaped reflector.

When ultra-violet radiators are suspended in a room thought should be given to the fact that objects surrounding them do cast shadows and restrict the spread of radiation, hence the radius of action.

Open air installations: It is a principle of open air deodorizing installations that a marginal area of at least 2 m (7 ft) width should be considered requiring irradiation in addition to the actual source of malodour, such as a waste heap of material, or a lagoon, an open tank, and similar objects.

The ultra-violet generators will be hung from masts or steel ropes across the area to be protected, and care should be taken to see that people do not get in such a position relative to the generators that they might suffer from conjunctivitis by looking inadvertently into the tubes. Fencing in the area and posting a notice board will satisfy legal requirements. Workers in such areas must be made to wear protective eyeshields which may be coloured or colourless and made of a plastic material which is opaque to the 2537 Å line, or are made of plain white glass.

To test a prospective material for its eye protecting qualities, an indicator is easily prepared. A small quantity of tungsten metasilicate sufficient to cover a sixpence is mixed with a suitable carrier which is transparent to the 2537 Å line. This may be either vinyl alcohol (water soluble) or one of the carriers used in the manufacture of fluorescent tubes for depositing and fixing the fluorescent powder.

A thin layer of tungsten metacilicate suspended in the carrier is applied by brushing or spraying to a suitable substrate, either metal, plastic, glass, or quartz. When exposed to the rays the powder will exhibit a strong green fluorescence. If the material intended for the protection of the eyes is placed between the source of radiation and the indicator, the green fluorescence should disappear completely if the screen is opaque to 2537 Å.

When the tungsten metasilicate has been applied to a piece of clear glass the opaqueness of the glass to ultra-violet rays can be demonstrated by irradiating the indicating powder directly when it will fluoresce, but will not be excited, i.e., remain white, when the radiation is made to strike the glass first. When quartz is used as a substrate fluorescence will be caused either way showing that quartz is transparent to ultra-violet radiation.

The green colour of fluorescence is very nearly identical with that caused by the green line of 5461 Å wavelength. This permits of measuring the fluoresence intensity either in terms relative to the intensity of a standard medium pressure (4–8 atm) mercury vapour discharge spectrum, or in terms of absolute output if the energy emitted by the mercury standard at 5461 Å, is known. This value can be taken from manufacturers' lists. The line is reliably separated by a liquid monochrome filter of tartrazine (a yellow pyrazolone dye). A quantity of the dye is dissolved in water until all blue spectral lines have disappeared. Check with a pocket visual spectroscope. Then add neodymium nitrate or didymium nitrate. The filter is stable and of excellent keeping quality (269). The thickness of the glass trough should be 50 mm.

Solid filters of photographic quality are commercially available for the same purpose, e.g., Chance-Pilkington type OY2; Ilford mercury green 807; Kodak 77 or 77A.

When destructive deodorization in the open is considered the direction of prevailing wind should be taken into account and the ultra-violet generators sited accordingly. In the leeward direction an ample marginal area will also be irradiated to catch any air currents drifting away from the source of odour.

Materials: Constructional materials must resist the corrosion of nascent oxygen, ozone, hydrogen sulphide, trimethylamine, and any other corrosive vapour that might occur in the atmosphere surrounding the ultra-violet generators.

Ventilation ducts may be made of standard materials except where there is a corrosion hazard. Irradiation chambers are preferably made of aluminium sheeting which can be anodized for increased corrosion resistance, but need not be so treated. The layer of natural aluminium oxide forming on the sheets will afford sufficient protection in most cases.

To prevent galvanic corrosion caused by contact between dissimilar metals in humid atmospheres, structural parts, bolts, nuts, washers, etc., should be of the same material, thus: aluminium–aluminium; steel–steel. Aluminium bolts may be replaced by cadmium-plated steel bolts. Brass, copper, or 18/8 stainless steel should not be used with aluminium under conditions of ultra-violet irradiation. Contact with steel sheets should also be avoided as far as possible, unless protected, e.g., by neoprene coatings.

Of plastics, polyvinylchloride, polychloroprene, and copolymers of vinylidene fluoride plus hexafluoropropylene, or chlorosulphonated polyethylene, give excellent service and are reliable for many years under the above conditions.

Protective coatings are useful as a final surface treatment. The following materials have been arranged in order of merit with regard to the usual combination of atmospheric hazards occuring in places where destructive deodorization is applied.

Table 50   *Anti-corrosive protective coatings* (60)
arranged in order of merit

Neoprene
Vinyl
Epoxy resins
Chlorinated rubber
Polystyrenes, mixed
Furans
Phenolic resins
Alkyds
Asphalt
Oil paint

41 Deodorization by change of process

The reduction of smell by a change in the manufacturing procedure may be sought or planned for economic or for purely technical reasons. In either case, research will play a leading part as a means of creating an industrial innovation which may rejuvenate not only the section concerned, but the entire works (226).

To economize on waste, the production process can be altered in such a way that it produces less waste; or it can be modified to produce a different type of waste which will lend itself to commercial use; or the waste can become a raw material for a new process.

The diagrams show a representation of various modes of waste control. The complete re-use of waste in the preparation of basic or raw material is illustrated in Figure 48(a). This scheme is applicable to many processes, for instance the one shown in diagram Figure 48(b). The general representation of what might be termed a closed circuit process, is shown in Figure 48(c), whereas an open circuit process is intimated in Figure 48(d).

Figure 48(b) is characteristic of those processes which are based on little or no chemical change. They are physical processes, and the matter appears as the finished product, or as waste. Figure 48(a) represents a process involving a great deal of chemistry. In fact, it is a highly complex process transforming a plant product (grass) into an animal product (milk, body building substances). It is a process which links plant metabolism with animal metabolism. The waste material (dung) is of the vegetable type and, therefore, directly returnable to the raw material (grass). But the circuit shown in Figure 48(d) is more representative of modern production, a characteristic of which is the complexity of the manufacturing process.

There are few production processes which do not produce some waste. This may be fundamentally of the same nature as the raw material to which it is usually returned for recovery. If the waste is different, it may

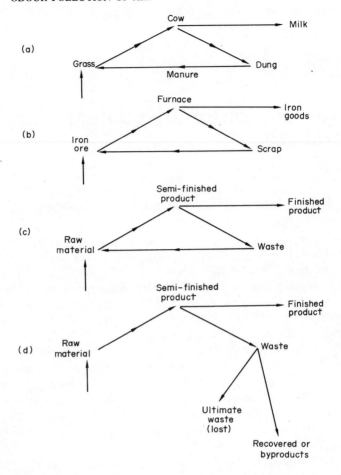

Figure 48   Some possible flow sheets.
(a) Closed circuit representing a biological process.
(b) Closed circuit as above, but of an industrial process.
(c) Closed circuit of an industrial process.
(d) Open circuit (linear) of an industrial process.

have a direct application, such as slag for bricks and road making, or it may require further processing in order to become a useful by-product, for instance petroleum and coal processing by-products. There are waste materials which have not found an application yet, and are therefore classed as total waste and uneconomical, for instance, smoke. The processing of pitchblende for the production of radium is an instance where the waste is 10^6–10^9 times the bulk of the purified end product; yet the intrinsic value, as against the economic value, is so high that this does not enter into consideration other than as a technological factor.

Changing a production process as a means of fighting air pollution is not a method readily appreciated by industry. When it can be shown that such a change is not only for the sake of making waste less offensive, but that it would turn waste into a saleable or otherwise useful by-product, the manufacturer will certainly be interested. A few instances are selected at random.

In the production of chlorinated hydrocarbons one-half of the chlorine supplied is transformed into hydrogen chloride $CH_4 + 6Cl \rightarrow CHCl_3 + 3HCl$ which could be termed a useful by-product if there were a market for it. As things were, this hydrochloric acid was sheer waste. The hydrochloric acid is now split by electrolysis yielding chlorine which is recirculated into the primary process of chlorination, and hydrogen which may be allowed to go to waste, or can be re-used in processes of hydrogenation. There is nothing new in regenerating chlorine to the turn of 100 t/day. In addition, a process using hydrogen chloride directly has been introduced, yielding another chlorinated methane, $HCl + CH_3OH \rightarrow CH_3Cl + H_2O$, which is the starting point in the production of carbon tetrachloride.

Atmospheric pollution from paint-finishing processes is dangerous not only because of the offensive strong smell of the solvents, but also because these are a real hazard to health caused by the discharge of organic binder and resin. Apart from standard methods of scrubbing, it is suggested that the true solution is the changeover from solvent-based paints to the equivalent water-based (emulsion) paints (176). Regulations controlling the emission of solvents in Los Angeles County (Rule 66) and the Bay Area (Regulation 3) especially merit study (244). See also Table 27.

Sulphuric acid is produced with an efficiency of 97·5%. The double contact process raises the efficiency to 99·5%, so that waste, SO_2, is reduced from 2·5% to 0·5% of the total tonnage produced (168). The residual sulphur dioxide is discharged from a tall chimney and sufficiently diluted to remain below osmic threshold. A process using an adsorption technique for the forming of sulphuric acid *in situ* on the adsorbant has been designed by Reinluft GmbH, (Clean Air Ltd), of Essen. Sulphur oxides are removed from flue gas by means of a two-stage adsorber. Low-temperature coke is used instead of the higher priced activated carbon (245). Another method is described in British Patent 1 090 306 (1967). The waste gases are passed through carbon adsorbers which were activated by impregnation with $0.1-0.15\%$ of ammonium iodides or ammonium iodates, or of an alkali metal. Coke may also be used at 100-160°C. The adsorbants can be reactivated and the sulphuric acid recovered by washing with water. Activated carbon beds achieved 77% desulphurization of flue gas which contained 0.3 vol. % SO_2, 3 vol. % O, 6 vol. % H_2O at 120°C after 30 min, and 39% desulphurization after 15 hours. Impregnating with 0.1% ammonium iodide, NH_4I, the absorption was 100% after 30 min, and 46% after 15 hours.

A recently developed method for the removal of hydrogen sulphide from gases is described in the Czech Patent 121 732 (1967). Hydrogen sulphide in a concentration of between 5 and 6 g/m^3 (5-6 mg/l) was removed by passing the gas through a layer of iron shavings which were continuously sprayed at the rate of 1:50 000 to 1:100 000 with a 10% solution of the sodium salt of bis(β-hydroxyethyl)glycine, or diethylene-triaminopentaacetic acid. Other solutions used contain β-hydroxymethyli-minodiacetic acid, triethanolamine, or bis(hydroxymethyl)glycine. A contact time from 3-5 sec is required. The solutions are all oxygen carriers.

Where the concentration of sulphur dioxide in air is small, mostly less than half the conventional concentration of between 8 and 12% SO_2, as in waste gases from smelting plant, a special process designed by KENNECOTT COPPER (169) will give high efficiency yields. Highest economy is aimed at by using almost exclusively special anticorrosion alloys and PTFE linings, pipes, and duct work. Efficient sulphur trioxide absorbers are supplemented by two-stage PTFE mesh mist eliminators.

One of the main sources of sulphurous fumes is the sulphur contained in coal. Another is the sulphur in fuel oil. Perhaps one of the most outstanding contributions to reducing osmogenic air pollution by change of production processes is that of the American oil industry. Although domestic crude oil has increased from 1950-1965 by 58% (from 2.1 to 3.3 billion (10^9) barrels a year), the percentage of total sulphur in crude which remains in fuel products decreased from 54.3% (1950) to 29.7% (1965). To absorb the sulphur dioxide in flue gas from these sources, the Japanese Mitsubishi Heavy Industries Company (170) uses activated manganese oxide, and a similar process using manganese is described in *Chem. Age* 84 (1960) 624. The Hitachi Company uses activated carbon as an adsorbing agent, and follows up this stage by a bath of caustic soda. The Mitsubishi process produces ammonium sulphate as a saleable by-product, and the Hitachi process sulphuric acid. The high yield, which is technically satisfactory, becomes an economic proposition especially where the oil-fired boilers are situated near a chemical works using either of the products.

The manganese dioxide (Mitsubishi) process may appear a mixed blessing in practice. Processes using particulate solids for the removal of gases from the air suffer from a limited mass transfer rate from gas to solid surface. The number of particles carried per unit volume cannot be increased beyond a certain proportion in the attempt to improve the mass transfer quantitatively, as the danger of precipitation grows with the concentration c according to $Dp \sim c^m$ where $m > 1$. In general, the contact time required for a 90 % removal of the solids from the air stream after purification would be impracticable (202). Although the low operating cost of the Mitsubishi process is attractive—to remove 90 % of the sulphur dioxide only \$1.10 would be added to the cost of each metric ton of fuel, and the process would also show a profit of \$32 for each ton of ammonium sulphate recovered—it would add a great load of discharged manganese dioxide to the amount of solid particulate matter already polluting the air.

It has been pointed out that even if 99·97 % of the manganese dioxide is used up, an industrial plant might easily release 50–60 t in the course of a year. It is extremely doubtful whether this high efficiency could be reached, or maintained, and the dust load would rise correspondingly. Suzuki (203) has shown—with no reference to the Mitsubishi process—that air pollution in Japan is characterized by the amount and concentration of sulphur dioxide and floating dust. The Yamaguchi Medical School points to the direct ratio of pollutant concentration and death rate. The 'Yokkaichi asthma' is an ailment caused by air pollution, and is characteristic of the Yokkaichi District. The disease becomes especially serious when the wind velocity is greater than 5 m/sec which causes a high concentration of fumes and dust. Relevant data are given by Imai, and the aetiology is discussed (243).

It is precisely in this locality that the Mitsubishi pilot plant will be operating. The Company is confident of success, and points out that only 25 % of the flue gases will be treated with manganese dioxide. Whilst the figure of some 50–60 t of solids per annum added to the dust load over Yokkaichi sound a formidable hazard, this should be seen in the light of total solids in the air. It is considerably less disquieting if expressed, as a result of test measurements, as an impurity added at the rate of 0·07 μg of manganese dioxide per m^3 of air. For this truly (?) negligible increase, nearly all of the sulphur dioxide emitted by the pilot plant, will be abated.

Although it may not appear to be an economic proposition at the moment, electrostatic precipitation offers a means of preventing the discharge of manganese dioxide.

A new process for the removal of sulphur dioxide from gasified coal or heavy fuel oil uses dolomite at high temperatures. The spent dolomite is regenerated with steam and carbon dioxide yielding sulphur as a valuable by-product (194). Hydrogen sulphide is removed from coke oven gas by means of ammonia. The liquor is then regenerated and the hydrogen sulphide recovered (195). A similar method uses limestone dolomite, a calcium magnesium carbonate, $CaMg(CO_3)_2$ (246).

The result of these and similar efforts is a steady decrease of sulphur in air. Even in the face of an ever increasing use of natural gas and petrol, there is less sulphurous air pollution (171). Sulphur dioxide in the liquid state from digesters used in the production of rayon pulp from wood chips can be recovered, for instance, by the CELLECO-SOMER process (188), and converted to sulphuric acid, or by catalytic conversion when it is a component of flue gas (189), or by a modified FOSTER WHEELER process when it is carried by smelter gas (190).

Not only production processes are being reviewed in the attempt to reduce pollution. Engines consuming solid or liquid fuel are investigated, too. At the Philips Works, Eindhoven, the Sterling engine was constructed; it causes considerably less concentration of exhaust gases than comparable petrol or diesel engines. The Sterling engine's outstanding feature is continuous, as against the discontinuous, combustion on which principle all other internal combustion engines are built. The walls of the combustion chamber remain hot enough all the time for the fuel to ignite spontaneously. The amounts of CO and hydrocarbons escaping to air are low, and the toxic lead compounds needed in petrol engine fuel can be eliminated.

Another project concerns the type of energy generation. Fossil fuels always produce a large amount of toxic, or dangerous, or offensively smelling waste gases. It has been suggested (172) that fossil fuels should not be burnt directly, but be used as raw materials for the synthesis of ammonia which can be termed a 'clean' fuel in the sense that its combustion does not give rise to any waste products. The fundamental equation is $4NH_3 + 3O_2 \rightarrow 6H_2O + 2N_2$. Complete combustion of ammonia in air yields pure water and nitrogen. The process is particularly attractive where there are large quantities of ammonia available as a waste product, as from nitric acid plant (175).

The various processes of producing viscose have always suffered under the heavy burden of hydrogen sulphide and carbon disulphide. Several variations of the process have been tried with different degrees of success. The most recent innovation relates to a new process for the purification of the exhaust air (177). Hydrogen sulphide is contained in the waste air extracted from the viscose producing machinery, while carbon disulphide is the solvent needed in the process. Hydrogen sulphide is formed by secondary reactions during the spinning and curing of the viscose filament. Beds of activated carbon are used for oxidizing the hydrogen sulphide, the fundamental process using nascent oxygen

$$H_2S + O \rightarrow H_2O + S \quad (+222 \text{ kJ})$$

Higher oxidation products are also formed, especially sulphuric acid

$$H_2S + 2O_2 \rightarrow H_2SO_4 \quad (+792 \text{ kJ})$$

All processes are exothermic. The generation of sulphuric acid is not a desirable phenomenon as it causes the carbon beds to get too hot. This may cause ignition of the carbon disulphide.

The PINTSCH-BAMAG process uses a new kind of activated carbon (cf. Section 31) which acts as a catalyst performing the oxidation of hydrogen sulphide to sulphur and water. Thus, only a small quantity of H_2S entering the carbon beds will be transformed into higher oxidation products. This eliminates all complications previously caused by high temperatures. The flow sheet, Figure 49, shows how the polluted air, after being exhausted from the viscose production plant, passes through a set of adsorbers operating in parallel. Hydrogen sulphide is catalysed to sulphur and water, and carbon disulphide is adsorbed on the activated carbon. The air, now freed from pollution, is released to atmosphere without further treatment. Only one half of the adsorbers is operative at any one time, the others standing by to take over when the first set reaches saturation. Regeneration is carried out *in situ* by reducing the oxygen content in the adsorber to below explosion limit. This is done by flushing with a gas poor in oxygen. Next, steam is passed through the beds desorbing the carbon disulphide. After condensing and cooling the H_2O-CS_2 mixture, the two ingredients are separated and the CS_2 is used for extracting the sulphur from the activated carbon. The slurry is then heated when the CS_2 evaporates and, finally, is returned to the process.

Another process removes hydrogen sulphide and carbon disulphide simultaneously in one plant which is of great economic importance to the viscose industry (196).

In the production of nylon, there are several methods of producing caprolactam, the raw material for nylon 6. The basic materials for caprolactam are aromatics (benzene, cyclohexane, phenol, toluene) and non-aromatics (acetylene, ethylene, furfural). A by-product is ammonium sulphate, and a highly offensive smell if the caprolactam, an amine, leaks during production. The process developed by the Toyo Rayon Company of Japan (178) uses the principle of photonitrosation of cyclohexane

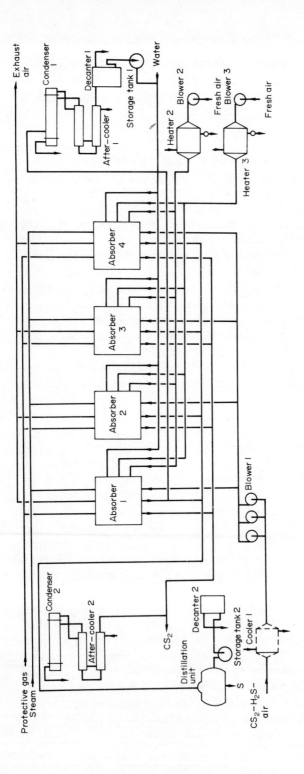

Figure 49 Flow sheet of a CS_2–H_2S recovery plant, system Pintsch–Bamag. The polluted air is cooled (cooler 1) and distributed (blower 1) to four absorbers into which a protective neutral gas and steam are admitted. The collected stream then goes to the distillation unit and, via a pump, to decanter 2 and thence to storage tank 2. Fresh air is blown via heaters 2 and 3 into the four absorber units, together with water. The liquid from the first absorber goes through condenser 1 and after-cooler 1 to decanter 1, whence it flows to storage tank 1 and is then recirculated via the water system. S = sulphur outlet.

(British Patents 1 010 151, 1 017 242, 1 066 114), and yields not only the cheapest ε-caprolactam by the fastest route, but also produces high quality ammonium sulphate which, in contrast to the by-product of other caprolactam processes, is produced only in half the quantities and can be sold directly as fertilizer. ε-caprolactam is

$$CH_2-(CH_2)_4-C{=}O$$
$$\lfloor\underline{\quad\quad} NH \underline{\quad\quad}\rfloor$$

The olfactory burden is not always negligible. In processing plants concentrations of caprolactam in air are normally between 2–40 mg/m^3.

The determination of ε-caprolactam in air has been described by GHERSIN (197). It is based on the transformation of caprolactam to hydroxamic acid, which contains the monovalent radical

$$-C\underset{NH.OH}{\overset{O}{\diagup\!\!\diagup}}$$

The latter is detected at concentrations of 4–80 mg/m^3, the sensitivity being 0·1 mg per 4 cm^3 of the sample solution.

An interesting step in the reduction of air pollution in cities has been proposed by the planning authority in Moscow. Not only has a ban been put on new factories in Moscow, but also about 200 works will be moved outside the city boundaries, and 340 of the remaining factories are to change the production processes so as not to cause air pollution. About 4000 air purification plants have been installed already in workshops and factories throughout Moscow (173).

An instance of a process material being substituted by another one to comply with anti-pollution regulations arose from the Los Angeles Rule 66 which prohibits the uncontrolled use of trichloroethylene, C_2HCl_3 by requiring that users install vapour emission control equipment. This caused a group of interested producers to propose that trichloroethylene be exempted from the rule. Their plea failed, not only because of considerations of public health, but also because another manufacturer has developed trichloroethane as a rule-66-exempt substitute (318, 319).

Part V

PREVENTION OF
ODOUR POLLUTION

42 Town planning

If war destruction had not necessitated rebuilding on an unprecedented scale, the rapidly expanding population would have presented the same problems. Building soon took up every acre available within, then around, a township, and houses mushroomed in areas where, before the war, only sewage treatment works and other offensive trades, were doing their part to keep society on a sanitary footing.

To guide and control the building of housing estates or whole districts in UK, Town and Country Planning Acts were introduced and regulations laid down in the various Statutory Regulations relating to the Acts. A whole string of New Towns was designed and built, with defined industrial, commercial, and residential areas. Smoke and soot and dust were banned from these New Towns, and so was noise. Even the industrial districts had nothing of the older conception of 'industry' about them. They were clean, and almost noiseless.

Yet, there was, and still is, one thing which has not been controlled to any extent—smell. Polluted air is no respecter of borders or areas or designations, Figure 50. New towns and old still suffer from disagreeable odours, if not to the same extent as in the past.

It is clear that Town and Country Planning is yet in its infancy because it has remained a national, rather than becoming an international, concern; because not sufficient experience is available to guide large-scale planning, and because many of the new towns soon had to look for satellite towns to accommodate their 'overspill' of people.

Only the aspects of air pollution by smell can be considered here with regard to the planning and building of new towns. The sectioning off of certain areas for different grades of occupation is not the only fundamental principle of good planning. Meteorological, topographic, orographic, and hydrological factors must be considered. An earth satellite took pictures of industrial haze over Houston, Texas covering an area of 2600

272

Figure 50   Change in wind direction with height. A change in stack height may change the location of pollution. The smoke trails were released during a temperature inversion at 50, 150 and 350 ft (15, 45 and 100 m) above ground. (*Courtesy Brookhaven National Laboratory*)

square miles (217). That is the total area of Devonshire. A spacecraft picked up a forest fire producing a smoke plume 4 miles wide and 65 miles long. It was trapped under an inversion and was being transported by the local wind flow pattern some 550–600 miles over land into the Gulf of Mexico.

Such cases surely demonstrate the need to include in the planning of new settlements much larger areas than just the few square miles on which houses will be erected. Economic and political factors will affect the picture. Meteorological data (219) can be usefully applied to the solution of air pollution problems of all kinds, but, as the International Clean Air Congress confirmed in 1966, only the drastic reduction of emission will finally solve the problem (218).

Whilst it will always remain a difficult task to site industrial and residential areas, the various solutions proposed or executed to date are far from ideal. Industrial estates should be far from residential areas because of air pollution by smoke, noise, and smell. But people working in these factories should live as close as possible to them to reduce transport problems to a minimum. These antagonistic and yet closely interdependent stipulations are always difficult of fulfilment. It is in circumstances like these that the study scheme of the Department of Health, Education, and Welfare (HEW) in the USA will give valuable guidance and evidence of the accommodative faculties of man with regard to changes in his environment. The cooperation of industrial planners is of the highest value and should be sought by town and country planning officials.

An instance is the building of the new oil refineries in Benicia, California. The company building the plant has caused a detailed study to be made of the environment. Data thus gained were used in the design of the refinery including the main stack. Within a radius of 7 km (4 miles), hundreds of air samples and specimens of soil, vegetation, and water were collected and analyzed, and a study was made of the meteorological conditions. Typical data characterizing the purity of air and water and soil before the construction of the refinery will be used as a reference level to guide the engineers during the working of the refinery. A point has been made and laid down as a regulation that the 'as before' level must on no account be exceeded during operation. Three monitoring stations are distributed through and around the refinery to keep watch on pollution (281).

A review by EFFENBERGER (220) of air pollution and its control with respect to the development of existing towns and the planning of new townships contains a great deal of interesting matter.

The United States National Centre for Air Pollution Control, an establishment of the Department of Health, Education, and Welfare (HEW), has instituted a nation-wide research into the social and economic impact of odour (221). This will reveal the main sources and geographic distribution of odour problems; it will measure by an exactly specified method the social impact of odours of all types on the various communities, and will try to determine how long it takes before an odour, irrespective of its character, becomes acceptable to the community. It is also intended to develop a technique which will permit of assessing odour impact rapidly and effectively.

Perhaps a little fantastic, but certainly not utopian, is the concept of building a new city, all-enclosed by a huge plastic dome transparent to light and health-giving ultra-violet rays, and constantly air-conditioned and kept free from microbes. Whatever the thoughts of critics, or of people who might be invited to live in such a city, there is one excellent feature about it: all services, and all service traffic, would be located underground so that air pollution would be non-existent (174).

43 The odourless factory

It is convenient to discuss this subject under two headings: existing buildings, and new buildings. Although the target is the same in either case, the approach to achieving this end is different in many details.

With *new buildings* the position of the site with regard to residential districts and prevailing wind should be studied carefully. The surveyor of the Local Council will usually be in a position to advise on this. If it is possible to site the new factory in the leeward of the residential district, half the battle is won. If the site is to windward of the residential area a full programme of preventive measures must be accepted.

Local climate and meteorological conditions essentially affect designs and decisions regarding chimneys, vent shafts, and similar structures. Normal atmospheric air flow patterns are governed by natural topography and the general principles of physics. Air is transparent to luminous and infra-red radiation, but absorbs, and is heated by, longer infra-red rays such as are emitted by the surface of the earth, which absorbs incident solar radiation. Because the earth acts as a (secondary) source of thermal rays, the air closest to it is warmest. Air temperature is, thus normally inversely proportional to altitude, and temperature lapse is normally a straight-line graph.

If the ground loses a great amount of heat by radiation over a restricted surface area, the temperature of the air closest to this area will drop below the temperature of higher strata of mostly unpolluted air. As it is the lowest layers of air which are polluted most, the higher strata are transparent to the radiation emitted by the lower-lying polluted air. This results in a quick and abrupt temperature drop so that the cooled air forms a 'bubble' resting on the ground. The air temperature above the bubble is higher than the bubble causing a temperature inversion.

There is hardly any air movement in an inversion bubble, and conditions are stable for long periods, even several days at a time. There are no convection currents in a bubble, and any discharges into the air in that bubble will drop to the ground. Cold air follows closely the contour of the ground affected by the inversion, and the density of the cold air causes it not to rise. Conditions in an inversion bubble are quiescent and stationary. The bubble may extend in a horizontal direction from a few yards to several miles, the height also varying between wide limits. Toxic fumes from chimneys discharging into the bubble act as a lethal respiratory poison, and smells trapped in an inversion may not be deadly, but yet affect people living in the bubble most seriously. With regard to air pollution the most direct and immediate phenomenon characterizing an inversion is the steady rise in pollutant concentration which, in the absence of dispersion, is maximum at ground level. Meteorological information can be obtained in many cases locally or, in England, from the Meteorological Office, Headstone Drive, Harrow, Middlesex.

Local Authorities require, as a rule, solitary works, such as sewage treatment plants, to be screened by tall trees. Obviously, the presence of sewage treatment plant is more offensive through the smell it produces than through the sight the works offer, and rows of tall poplar trees cannot retain the smell within the confines of the works. On the contrary, these trees like any other tall objects in the wake of odoriferous air currents act, after a time, as secondary sources of odour.

Where water is used as a coolant, especially in nozzle scrubbers, cooling towers and wherever the water makes intimate contact with polluted air, the used water will itself become a source of offensive smells and discharge into the sewers may not be permitted without prior treatment. But such treatment often puts the pollutant back into the air either directly, or indirectly.

Machinery producing or emitting odoriferous air should be kept confined to certain small sections if the manufacturing process will permit it. The entire section should then be made airtight and ventilated by extracting the air through ducts, creating in the section a negative atmospheric pressure. If it is a single machine this must be totally enclosed and continually ventilated through ducting. It is essential to use the services of the ventilating engineer during the early design stages. Polluted process air is kept separate from the non-polluted air so that the air volume undergoing treatment is kept to a minimum.

A general layout plan will be modelled on the scheme in Figure 51. It is important to use only exhaust fans, not pressure fans. This will prevent leakages of polluted air into the environment of their sources.

Good maintenance is essential to the proper working of an air treatment plant, and frequent and regular inspection of the treatment chamber and

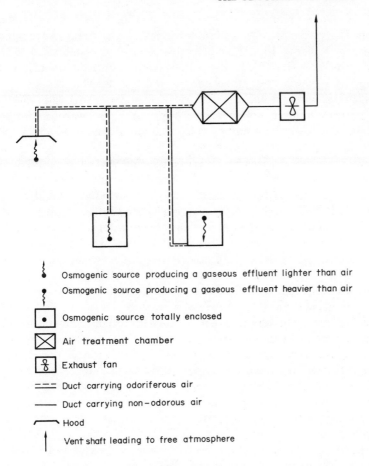

Osmogenic source producing a gaseous effluent lighter than air

Osmogenic source producing a gaseous effluent heavier than air

Osmogenic source totally enclosed

Air treatment chamber

Exhaust fan

Duct carrying odoriferous air

Duct carrying non-odorous air

Hood

Vent shaft leading to free atmosphere

Figure 51   General scheme for the treatment of polluted air.

its equipment must be carried out. The most suitable scheme for the treatment of smell polluted air is destructive deodorization (Section 36). This is the only method not raising any problems about repolluting the atmosphere.

With *old buildings*, conditions are quite different. When a factory has been housed in an old building for some time the osmogenic nuisance has become almost an institution. Suddenly, it may be challenged in Court.

It is usually after the abatement notice has been served on the management, that the thinking about methods will start. The principle of separating process air from general air is almost impossible of achievement as neither layout nor type of machinery is conducive to that end.

The volume of air requiring treatment is, therefore, large and the work to be done fairly expensive. An additional difficulty is often presented by the state of the building. In a great many instances of old established works carrying on offensive trades, the roof is simply an iron structure with corrugated asbestos sheeting fastened to the trusses. The sheets usually sit with their corrugations at the eaves of the supporting walls without the corrugations ever having been stopped off. The warm air in the building rises to the roof and escapes in a continuous stream of smells and steam through the corrugations.

Windows and doors are kept wide open during the summer, and doors even during the winter as they, usually, are the loading gates at the same time. Any attempt by the management to keep windows closed will be unsuccessful, because the people working in the foul atmosphere of the place insist on what little fresh air may be able to enter the building by the numerous cracks and leaks, or through doors and windows, and cross-draught helps. That this 'fresh' air may be as polluted as the air within the factory, seems not to matter. Relief is purely psychological.

For the environment of the factory, these open doors and windows are one of the most powerful sources of obnoxious odours. From practical experience, it must be admitted that, unless a thorough—and expensive—scheme of ducting and building repairs is carried out the nuisance will not be reduced. Half-measures are just a waste of money; they achieve nothing. In these old buildings, where such trades as the production of fertilizer from animal waste, and others producing strongly smelling dust, have been carried on for some time, every crack in the wall, every brick jutting out from the wall, and any bit of machinery, roof beams, and others, are thickly covered with the offensively smelling dust. This is a most powerful source of odour day and night, weekdays and Sundays, regardless of whether the factory is working or not.

A general scheme for alleviating such a smell nuisance will comprise, at least, the following points:

(A)

(1) Repair the fabric of the building; seal all cracks.
(2) Control the opening of doors and windows.
(3) Clean out all accumulated dust.
(4) degrease the floor and lower parts of the walls.
(5) If at all possible, make the shell of the building airtight.

(B)

(6) Re-arrange the machinery to avoid large distances between units.
(7) Enclose each source of odour.

(8) If that is not possible, take measures to prevent any odours released during charging or discharging periods, from spreading uncontrolled or untreated.

(9) Install positive control of air movement.

(10) Make sure that the entire place is at a slight negative pressure; this will prevent the uncontrolled spreading of foul air to the environment.

(C)

(11) Install a method of air deodorization appropriate to the manufacturing process and operating conditions.

(12) Use an expert's advice; it is cheaper in the long run.

A questionnaire is useful as an aide-mémoire to odour problems of manufacturers who are a long distance away from consulting engineers, or do not wish for a visit from an expert in the early stages of their inquiry. It should elicit all important details and assist the engineer to visualize the factory and its problems with fair accuracy.

QUESTIONNAIRE

(Underline italic words applicable)

NAME
Address
Telephone
Telex
Person to contact

TYPE OF PRODUCT
Type of process
Odour caused by: *product   by-product   waste*
Condition of: *product   by-product   waste*
     solid:
     liquid:
     gaseous:
Is waste to be discarded: *yes   no*
Pollution of: *environmental air   process air   liquid effluent   gaseous effluent*
Source of pollution is: *singular   several   general*
Is a ducting system installed: *yes   no*
Cross-sectional dimensions of ducts:
Volume of air: ft^3/min (or m^3/h):
Volume of rooms so ventilated: ft^3 (m^3), or dimensions:

CONDITION OF BUILDING:
 roof in good order: *yes no*
 are windows kept: *open closed*
 are gates (doors) kept: *open closed*
 are walls and roof trusses: *clean covered with process dust*
 is the floor: *clean dirty greasy*

USE OF BUILDING: *production storage of: raw materials finished goods*
Loading of vehicles in: *the open semi-open bays totally enclosed bays*

IS THE AIR RELEASED INTO THE OPEN: *treated untreated*
What kind of air treatment is used:
Is the treatment satisfactory: *yes no*
Is liquid effluent discharged to: *sewers rivers ditch tanks lagoons*
Is it treated before discharge: *yes no*
What kind of effluent treatment is used:
Is this treatment satisfactory: *yes no*

IS THE AIR DUST-LADEN: *yes no*
Is a cyclone used: *yes no*
Are there traces of dust around the air outlet on the roof: *yes no*
Is the foul air released through a chimney stack: *yes no*
 and together with flue gas: *yes no*
Is (white) steam mixed with the foul air: *yes no*
Temperature of discharged air: °C or °F

ENVIRONMENT:
Plan of environment (scale to   ): *enclosed herewith to follow not available*
Approximate distance factory to residential district: *miles* or *km*
Direction of prevailing winds:
Are there tall buildings or high trees near the factory *yes no*
(Indicate their situation in plan with distance)

44 Air pollution prevention service

The severity of air pollution caused by a single plant depends on both volume of production and kind of process, though there is no direct relationship between the two. A process might cause a high degree of air pollution though the output of the factory might be small.

Where there are many and varied processes, operating on a non-stop basis as in large chemical or pharmaceutical works, foundries, etc. there is an urgent need for an anti-air pollution officer or a whole section devoted to this effort. This is by no means an administrative position, but one which requires a highly trained engineer thoroughly familiar with the layout and function of the whole plant and every single section of it. He must also know of all legal requirements, rules, regulations, and local by-laws, concerning his plant. Large works usually have already a safety officer who is responsible for the prevention of injury to workers and damage to machinery. He should not be burdened with the additional duties of air pollution abatement or prevention, because the aims and methods involved are quite dissimilar. The air pollution prevention officer will also concern himself with the disposal of liquid waste and its pre-treatment before discharge to the sewers.

Fundamentally, this officer is concerned with emissions from machinery within the confines of the works whether within buildings or in the open, and with immissions, i.e., the concentrations of offensive waste (solid, liquid, or gaseous) on or near the ground within the area of the works; but also taking in the immediate, perhaps even the more distant, environment depending on the type of pollution caused. He will also consider topical, meteorological, climatic, and a whole host of other characteristics. Once the general geographical and climatological conditions around the plant have been established and charted, the following routine is introduced:

(A) *Scope:*

(1) Immission: general evaluation of air pollution over the factory area and the immediate environment;

281

(2) Emission: measurement of pollutant concentrations in the lee of individual sources (chimneys, vent shafts);

(3) Emergency: sudden outbreaks of smoke, steam, odoriferous fumes, or other air pollutants must be immediately recognized, measured, and stopped.

(B) *Methods*

The following paragraphs are modelled on work systematically carried out at Farbwerke Hoechst AG, Frankfurt/Main, whose efforts, equipment, and instrumentation have been leading the way (166, 167). The most important single factor is the incorporation of the recommendations of the waste disposal group into any plans for alterations or new developments. This goes so far that no new capital is allocated unless the waste disposal group has recommended the production process as not contributing to the pollution of air and water. The adopted methods enable the daily output of 56×10^6 m^3 of waste gas to be kept under control, i.e., to be cleaned of dust and fumes; for the treatment of water there are 49 sedimentation lagoons and numerous chlorination tanks available.

The scheme is shown in Figure 52. Recording instrument stations are arranged on the periphery of the works in such a manner that each serves a definite section of the plant with regard to the direction of prevailing wind. Each station is a laboratory on wheels, Figures 53(a) and 53(b), with a working area of 2 m $\times$ 3 m (6 m^2) (about 7 ft $\times$ 10 ft). The laboratory is thermally insulated and air-conditioned. Air pipe lines are of glass, and ground spherical joints connect the various parts. Concentrations of SO_2, NO_2, CO_2, NH_3, Cl are continuously monitored as a routine. Dust is also evaluated.

Two mobile laboratories for the measurement of gas concentrations are built into vans and are intended for use in an emergency, be it a sudden outbreak of gas or fumes due to a breakdown in the works, or serious and massed complaints from any part of the environment. Additionally, one of these laboratories on wheels is fitted with a radio telephone transmitter and receiver for communication with the works.

The mobile laboratories are fitted with robust instruments which give results quickly. A small section of the general service is a simple weather station, to give readings of wind velocity, atmospheric pressure, relative humidity, duration of sunshine. It also has a rain gauge. Wind is measured 60 m (200 ft) above ground and the readings are transmitted to, and registered by, instruments in the laboratory where also the other data are recorded.

The concentrations of sulphur and nitrogen dioxides are evaluated every half hour, and their direction of drift and intensity are indicated by

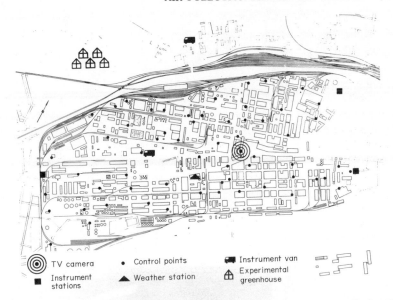

Figure 52 General plan of the air pollution control system at Farbwerke Hoechst A.G., Frankfurt/Main—Hoechst.

Figure 53(a)   Instrument van with experimental plants on the roof.

Figure 53(b) Instrument van, interior view.

arrows in a map which bears a direct relationship to the maximally accept-able concentrations of various pollutants in air. This information is passed to all managers of production departments.

The outstanding feature of this prevention service is a fully tran-sistorized television camera, Figure 54(a) with zoom lens having visual angles between 4° and 27°, magnification about 8:1, and an effective range of 1·5 km (1 mile). The camera is still useful, on a clear day, over 4 km (2·5 miles) which distance is, however, greater than maximum distance within the perimeter of the works. The movement of the camera and of the zoom lens is controlled from the laboratory where also the television receiver stands. Transmitter and screen are connected by 600 m (2000 ft) of coax cable. The camera turns full circle within 2 min. Displays on the screen can be photographed.

Figure 54(a)   TV camera for the observation of the air space above the works. The camera has a zoom lens (4 deg to 27 deg) covering the area of the works adequately. In clear weather, objects at the distant horizon (4 km = 2·5 miles) can be seen clearly.

Figure 54(b)   TV screen and control box. (*Courtesy Farbwerke Hoechst A.G.*)

A constant watch is kept on the screen to locate any sudden emissions anywhere in the works, Figure 54(b). The sensitivity of indicating or measuring instruments suffers from the common evil that immission concentrations are so low that instruments are no longer adequately sensitive and the investigator would have to rely on organoleptic methods. This unsatisfactory way of 'measuring through the nose' is obviated by instrumentally investigating the concentration at the point of emission and, using mathematical models for the dispersion and dilution of gas mixtures, from which is calculated the immission concentration to an acceptable degree of accuracy. Gas chromatography and flame ionization are the most sensitive detection methods. With a detector of the flame ionization type an organic pollutant gas at $0 \cdot 1$ mg/m^3 concentration can still be accurately determined.

Another branch of the air pollution prevention service is concerned with studying the effects of air pollutants on vegetation, for damaging or toxic waste effluents, may severely affect husbandry in the region.

The arrangement in Hoechst is quite simple. There are four groups of five small buildings of the greenhouse type. One building in each group is connected to a plastic pipe line carrying, in three instances, polluted air of exact chemical composition, the fourth carrying clean, i.e., unpolluted, air. Concentrations can be varied and are measured at each station by means of a portable instrument so that an average value can be established for each greenhouse. Effects of physiological development and condition, time of action, season, time of day when the pollutant is offered to the plants, and other characteristic data are carefully recorded. These investigations have helped to establish the accepted tolerance values of 0.5 mg/m^3 for sulphur dioxide (for constant pollution), and of 0.75 mg/m^3 SO_2 as a short-term value.

References

1 JOHN EVELYN, *Fumifugium*, 1661, reprinted by the National Society for Clean Air, London, 1961.
2 MCCORMICK, R. A. and J. H. LUDWIG, *Science* **156** (1967) 1358.
3 WENT, F. W., *Tellus* **18** (1966) 549–560.
4 *Clean Air Act*, 1956, 4 and 5 Eliz. 2, Ch. 52.
5 *Noise Abatement Act,* 1960, 8 and 9 Eliz. 2, Ch. 68.
6 BRITISH STANDARD SPECIFICATIONS:
    2740: 1956 *Simple Smoke Alarms and Alarm Metering Devices.*
    2741: 1957 *Recommendations for the Construction of Simple Smoke Viewers.*
    2742: 1958 *Notes on the Use of the Ringelmann Chart.*
    2742C: 1957 *Ringelmann Chart.*
    2811: 1957 *Smoke Density Indicators and Recorders.*
    1747: Parts 1–4: 1961 *Methods for the Measurement of Air Pollution.*
7 BRITISH STANDARD SPECIFICATIONS:
    4196: 1967 *Guide to the Selection of Methods of measuring Noise emitted by Machinery.*
    4198: 1967 *Method for calculating Loudness.*
8 *Petroleum Press Service*, **34** (April 1967) 145.
9 *The Town and Country Planning (Use Classes) Order* (HMSO, London, 1950).
10 ANON., *Environ. Sci. Technol.* **1** (1967) 192–196.
11 MCHUGH, E. W., *APCA* **17** (1967) 277–279.
12 LICHTENSTEIN, S., *Air Engng* **9** (1967) Nov. 12–15.
13 PARKER, A., *Chemistry in Britain* **3** (1967) 261–262.
14 ROHRMAN, F. A. *et al., Power* **111** (1967) May 82–83.
15 ANON., *Heat and Management, Tokyo* **15** (1966) July 7–18.
16 UNO *Yearbook* 1966, Geneva.
17 ANON., *Protection* **6** (1967) May 47–52.
18 HARTLEY, H., *New Scientist,* 28th May, 1964.
19 MONCRIEFF, R. W., *The Chemical Senses* (Leonard Hill, London, 1951).
20 JONES, R. and P. PYMAN, *J. Chem. Soc.* **127** (1925) 2588–2598.
21 AMOORE, J. E., *Perfumer and Essential Oil Rec.* **43** (1952) 321.
22 AMOORE, J. E., in Recent Advances in Odor: Theory, Measurement, and Control, *Ann. N.Y. Acad. Sci.* **116** (1964) 457–476.
23 PASQUILL, F., *Meteorol. Mag. Lond.* **90** (1961) 33.
24 BRANCA, G., *Fumi Polveri, Milano* **1** (1967) 7–15.

287

25 BOSANQUET, in THRING, M. W., *Air Pollution* (Butterworth, London, 1957).
26 SUTTON, O. G., *Quart. J. Roy. Meteorol. Soc.* **73** (1947) 426.
27 *The Tall Stack,* National Coal Policy Conference, Inc., 1000 Sixteenth Street, N.W., Washington D.C. 20036.
28 CHALKER, W. R., *Petrochem. Engr* **39** (1967) May 35–38.
29 CLOSSON, J., *Revue ind. miner.* **49** (1967) 252–257.
30 CULKOWSKI, W. M., *Nucl. Safety* **8** (1967) 257–259.
31 AVOYAN, A. O., *Sb. Tr. Inst. Epidemol., Gig.Atm.S.S.R.* **4** (1965) 156–163.
32 GLÜCKAUF, E., *Compendium of Meteorology* (Amer. Meteorol. Soc., New York, 1951).
33 PANETH, C., *Quart. J. Roy. Meteorol. Soc.* **63** (1937) 436.
34 THOMPSON, R. CAMPBELL, *A Dictionary of Assyrian Chemistry and Geology,* p. 129 (Clarendon Press, Oxford, 1936).
35 BOSANQUET, C. H. and J. L. PEARSON, *Trans. Faraday Soc.* **32** (1936) 1249.
36 KLUG, W., *Beitr. Phys. Atmos. Frankfurt* **33** (1960) 101.
37 CATE, F. L., *Tappi* **36** (1953) 225.
38 BODURTHA, F. T., *JAPCA* **11** (1961) 431.
39 CROSSE, P. A., D. H. LUCAS and W. L. SNOWSILL, *J. Inst. Fuel* **34** (1961) 503.
40 CONNOR, W. D. and J. R. HODKINSON, *Public Health Service Publication* No. 999–AP–30 (Washington, 1967), 89 pp.
41 DU BOIS and DU BOIS, *Arch. Internal Med.* **17** (1916) 863–877 (or any textbook on Human Physiology).
42 NATIONAL BUREAU OF STANDARDS, *Technical News Bulletin* **47** (1963) 175–177.
43 SFORZOLINI, G. S. and M. MARIANI, *Boll. Soc. Ital. Biol. Sper.* **37** (1961) 766.
44 TURK, A. and C. J. D'ANGIO, *Paper* 61–13 to Air Poll. Control Assoc., 54th Ann. Meeting, 1961, New York.
45 SHERRINGTON, C. S., *The Integrative Action of the Nervous System,* p. 8 (Cambridge Univ. Press, 1947).
46 CROCKER, E. G. and L. F. HENDERSON, *Amer. Perfumer* **50** (1947) 164.
47 HENNING, H., *Der Geruch* (Leipzig, 1916).
48 ERB, R. C., *Osteopath. Digest* **3** (1927) 4.
49 ZWAARDEMAKER, H., *Die Physiologie des Geruches* (Engelmann, Leipzig, 1895).
50 BARNEBEY, H. L., *Heating, Piping, Air-Conditioning* **30** (1958) 153.
51 BRAIN, W. R., *Mind, Perception, and Science* (Blackwell, Oxford, 1951).
52 CALVERT, S., *Mach. Design* **40** (1968) 35.
53 DUCKWORTH, S. and E. KUPCHANKO, *JAPCA* **17** (1967) 379–383.
54 ROCHKIND, M. M., *Envir. Sci. Technol.* **1** (1967) 434–435.
55 ANON., *Science J.* **4** (1968) 25.
56 KEANE, J. R. and E. M. R. FISHER, *Report* AERE–R–5366 (Harwell, 1967).
57 FISCHOTTER, P., *Kontinuierliche Messung der Staub und Gas Emissionen, Essen* **71** (1965) 24–31.
58 LIDZEY, R. G. and F. M. LONGMAID, *Chem. Ind.* (1964) 150.
59 MAGOS, L., *Brit. J. Industr. Med.* **23** (1966) 230–236.
60 SUMMER, W., *Methods of Air Deodorization* (Elsevier, Amsterdam, 1963).
61 ARNOLD, C. and C. MENZEL, *Ber. deutsch. Chem. Ges.* **35** (1902) 329.
62 ARNOLD, C. and C. MENZEL, *Ber. deutsch. Chem. Ges.* **35** (1902) ii, 321.
63 WADELIN, C. W., *Anal. Chem.* **29** (1957) 441.
64 KAESS, G., *Australian J. Appl. Sci.* **7** (1956) 242.
65 ALLISON, A. R. *et al.,* US Patent No. 2, 849, 291 (1958).

66 BOVEE, H. H. and R. J. ROBINSON, *Anal. Chem.* **33** (1961) 1115.
67 BOWEN, G., in *Aviation Medicine*, Agardograph 25 (Pergamon Press, London, 1958).
68 DELMAN, A. D. *et al.*, *Advances in Chemistry*, Ser. 21: *Ozone Chemistry and Technology*, p. 119 (Amer. Chem. Soc., 1959).
69 EHMERT, A., *Advances in Chemistry*, Ser. 21, p. 128 (Amer. Chem. Soc., 1959).
70 REGENER, V. H., *Advances in Chemistry*, Ser. 21, p. 124 (Amer. Chem. Soc., 1959).
71 KOBAYASHI, J., M. KYOSUKA and H. MURAMATSU, *Pap. Meteorol. Geophys., Tokyo* **17** (1966) 76-96.
72 KOBAYASHI, J. and Y. TOYAMA, *Pap. Meteorol. Geophys., Tokyo* **17** (1966) 97-126.
73 GERMAN, A., A. M. PANOUSE-PERRIN and A. M. QUERO, *Ann. Pharm. Franç.* **25** (1967) 115-120.
74 NASH, T., *Atmosph. Environ.* **1** (1967) 679-687.
75 DEUTSCH, S., *JAPCA* **18** (1968) 78-83.
76 COHEN, I. R. and J. J. BUFALINI, *Envir. Sci. Techn.* **1** (1967) 1014.
77 BUKOLOV, I. E. and N. I. NECHIPORENKO, *Gig. Sanit.* **32** (1967) 68-69.
78 ROBBINS, R. C. *et al.*, *JAPCA* **18** (1968) 106-110.
79 GRIFFIN, A. E., *Water Works and Sewage* **80** (1933) 218.
80 GRUNE, W. N., *Industr. Water Wastes* **7** (1962) 29.
81 SMITH, A. F. *et al.*, *J. Appl. Chem.* **11** (1961) 317.
82 ADLEY, F. E. and C. P. SKILLERN, *Amer. Industr. Hyg. Assoc. J.* **19** (1958) 233.
83 HUGHES, E. E. and S. H. LIAS, *Anal. Chem.* **32** (1960) 707.
84 TURK, A., *Public Health Service Publication* No. 999-AP-32 (USA, 1967).
85 COORDINATING RESEARCH COUNCIL, *Report* No. 401, New York, Oct. 1966.
86 STOPPS, G. J. and M. MCLAUGHLIN, *Amer. Industr. Hyg. Assoc. J.* **28** (1967) 43.
87 DAVE, J. V. and C. L. MATEER, *J. Atmosph. Sci.* **24** (1967) 414-437.
88 SIANU, E. and C. RADULIAN, *Igiena, Bucharest* **15** (1966) 561-566.
89 LIECHTENSTEIN, S., *Air Engng* **9** (1967) 12.
90 PATTERSON, C. D. and G. H. KAMP, *Amer. Rev. Respir. Diseases* **95** (1967) 443.
91 ROHE, K. H., F. J. MONIG and K. BISA, *Beitr. Problem der Luftreinhaltung, Essen* **68** (1965) 25-34.
92 BOESEN, V., *West Magazine of the Los Angeles Times*, 16th Apr., 1967.
93 TARASOV, S. I. *et al.*, *Gig. Sanit.* **32** (1967) 52-56.
94 MUIR, D. C. F., *J. Appl. Physiol.* **23** (1967) 210-214.
95 WATSON, A. J. *et al.*, *Brit. J. Industr. Med.* **16** (1959) 274-285.
96 KILBURN, K. H., *Arch. Environ. Health* **14** (1967) 77-91.
97 BATTIGELLI, M. C. *et al.*, *Arch. Environ. Health* **12** (1966) 460-466.
98 BIERSTEKER, K. and H. DEGRAAF, *T. soc. Geneesk.* **45** (1967) 74-77.
99 FREY, J. W. and M. CORN, *Amer. Industr. Hyg. Assoc. J.* **28** (1967) 468-478.
100 HALDANE, J. S., *Methods of Air Analysis*, p. 148 (Charles Griffin, London, 1935).
101 PIPER, S., *Pflüger's Arch.* **124** (1908) 591.
102 PÖTZL, K. and R. REITER, *Aerosol Forsch., Stuttgart* **13** (1967) 372.
103 PÖTZL, K. and R. REITER, *Acta Albertina, Regensburg* **26** (1966) 67.
104 WORTH, J. J. B. *et al.*, *J. Geophys. Res.* **72** (1967) 2063-2068.
105 JUNGE, CHR. E., *US Government Res. Devel. Rep.* **67**(7) (1967) 6.
106 WIENER, H., *New York State J. Med.* **66** (1966) 3153-3170.
107 V. BEKESY, G., *J. Appl. Physiol.* **19** (1964) 369.
108 BRILL, A. A., *Psychoanalyt. Quarterly* **1** (1932) 7.

109 LAIRD, D. A., *J. Abnorm. Social Psychology* **29** (1935) 459
110 PERRY, J., *Our Polluted World* (Franklin Watts Inc., New York, 1967).
111 MANUFACTURING CHEMISTS' ASSOC. *Source Materials for Air Pollution Control Laws*, Chicago, 1967.
112 *Lutte contre la Pollution Atmosphérique. Législation en Allemagne, Angleterre, Belgique, États-Unis, France, Hollande, Italie, Suisse*. Centre Interprofessionnel Technique d'Études de la Pollution Atmosphérique (CITEPA), Paris, CA/DOC 107 (25), 1966.
113 SMITH, M. E., *Mech. Engng* **90** (Feb. 1968) 20–22.
114 BLOKKER, P. C., *Ingenieur, The Hague* **80** (March 1968) 25–32.
115 GENERAL ELECTRIC CO., *Prod. Engng* **39** (Feb. 1968) 22–23.
116 ANON., *Clean Air, Tokyo* **4** (Apr. 1966) 10–19.
117 GOLDENBERG, S., *Rock Prod.* **70** (June 1967) 88–89, 104.
118 ELDRIDGE, E. F., *Industrial Waste Treatment Practice* (McGraw-Hill, New York, 1942).
119 THELE, E. JR., *Oil and Gas J.* **59** (1961) 84.
120 FIRESTONE, J. P. *Corrosion* **16** (Dec. 1960) 9.
121 FØYN, E., in M. W. THRING (ed.), *Air Pollution* (Butterworth, London, 1957).
122 SVERDRUPP, H. H. *et al., The Oceans, their Physics, Chemistry, and General Biology* (Prentice Hall, N.Y., 1954).
123 ROSE, H. E. and P. H. YOUNG, *Proc. Inst. Mech. Engrs* **1B**, 1–12 (1952–53) 114.
124 LEVA, M., *Tower Packings and Packed Tower Design* (The US Stone Ware Company, 1951).
125 MEHTA, D. S. and S. CALVERT, *Envir. Sci. Technol.* **1** (1967) 325–331.
126 RANDALL, C. W. and J. O. LEDBETTER, *Amer. Industr. Hyg. Assoc. J.* **27** (1966) 506–519.
127 ŽIŽKA, J., *Czech. Heavy Industry* (Nov. 1967) 18–26.
128 PRAGUE RES. INST. CHEM. EQUIPM., *Brit. Chem. Engng* **13** (1968) 299.
129 DAVIES, C. N., *Proc. Inst. Mech. Engrs* **1B**, 1–12 (1952–1953) 185.
130 HEYWOOD, H., *Proc. Inst. Mech. Engrs* **1B**, 1–12 (1952–1953) 169.
131 PORATH, J. and P. FLODIN, *Nature* **183** (1959) 13th June.
132 HERSCH, CH. K., *Molecular Sieves* (Reinhold, New York, 1961).
133 GUSTAFSON, P. and S. H. SMITH, JR., *JAPCA* **7**, 3 (1961).
134 KUNIN, R. *et al., J. Amer. Chem. Soc.* **84** (1962) 305.
135 McBAIN, J. W., *Sorption of Gases by Solids* (Routledge, London, 1932).
136 APCA–TA 7 Odour and Gas Treatment Committee, Chairman H. L. Barnebey, *JAPCA* **15** (1965) 422.
137 HOFFMAN, G. *Gas und Wasserfach* **102** (1961) 166.
138 NICKLIN, T. and E. BRUNNER, *Gas World* **153** (1961) 744, 841.
139 HAINES, H., *Oil and Gas J.* **58** (1960) 165.
140 GOUCHARENKO, G. K. *et al., Trudy Khar'kov politekh. Inst. V.I. Lenina, Ser. Khim. Tekhnol.* **13** (1957) 53 (in Russian).
141 WIENER, H. B. and P. W. YOUNG, *J. Appl. Chem.* **8** (1958) 336.
142 HUETER, T. F. and R. H. BOLT, *Sonics* (Chapman & Hall, London, 1955).
143 LAHMANN, E., *Staub* **26** (Dec. 1966) 24–29.
144 DRATWA, H. *et al., Chemie-Ing. Technik* **39** (1967) 949–955.
145 SHAH, R. K. *et al., Nature* **218** (1968) 593.
146 RONALD, D., *Offensive Trades;* London 1935 (now out of print).

147 MORRIS, D., *The Naked Ape: A Zoologist's Study of the Human Animal* (Cape, London, 1967).
148 KLEIN, H., *CAV* (1967) 35–38; (1968) 37–40.
149 GOTHARD, S. A., *Public Cleansing J.* (1959) 503–518.
150 BUSFIELD, J. E., *Surv. Mun. Engr* (1966) 3rd Dec.
151 OGDEN, J. G., *Contamination Control* 7 (1968) 8th June.
152 BRIGGS, R. et al., *Analyst* **83** (1958) 304.
153 POMEROY, R. and F. D. BOWLUS, *Sewage Works J.* **18** (1946) 597.
154 MÜLLER, W. J., *Gas Wasserfach* **102** (1961) 986.
155 ELDRIDGE, E. F., *Industrial Waste Treatment Practice* (McGraw-Hill, New York, 1942).
156 CHAYEN, I. H., *J. Appl. Chem.* (1953) 3.
157 OTHMER, D. F., *Chem. Process Engng* **49** (1968) 109–112.
158 ANON., *Chem. Process Engng* **49** (1968) 102—103.
159 LOQUERCO, P. and W. J. STANLEY, *Air Engng* 9 (Nov. 1967) 22–29.
160 MAY, J., *Staub-Reinhaltung der Luft* **26** (1966) 385–389.
161 FAITH, W. D., *Air Engng* 10 (Feb. 1968) 16–17.
162 MANUFACTURING CHEMISTS' ASSOC., *Manual* W–1.
163 LEPKOWSKI, C. W., *Chem. Engng News* **46** (11th March, 1968) 8A–17A.
164 ANON., *Contamination Control* 7 (1968) 27–29.
165 BAKER, R. A., *J. Amer. Water Works Assoc.* **55** (1963) 913–916.
166 TROBISCH, K., *Zentralblatt Arbeitsmed. Arbeitsschutz* **14** (1964) 82–86.
167 TROBISCH, K., *Zentralblatt Arbeitsmed. Arbeitsschutz* **15** (1965) 205–209.
168 SCHAAFHAUSEN, J., *Allgem. Prakt. Chemie, Vienna* **18** (1967) 77–79.
169 COPPER, K., *Canad. Chem. Process* **51** (May 1967) 78.
170 ANON., *Power* **111** (March 1967) 93.
171 MALLATT, R. C., *Hydrocarb. Process.* **46** (Dec. 1967) 115–118.
172 GREEN, L., JR., *Science* **156** (1967) 1448–1450.
173 ANON., *Engineer, London* **224** (1967) 670.
174 BOURNE, H. K., *UK Scientific Mission Report (North America)* No. 68/19, issued by UKSM, Washington.
175 BAYLEY, R. W., *Effl. Water Treatm. J.* 7 (1967) 78–84, 154–161.
176 OLIVER, J., *Prod. Finish., Cincinnati* **31** (July 1967) 62–69.
177 PINTSCH-BAMAG, *Sulphur* No. 74 (1968) 37–38.
178 HULME, P. and P. E. TURNER, *Chem. Process Engng* **48** (Nov. 1967).
179 STANBRIDGE, H. H., *Water Waste Treatm.* 10 (July 1964) 82.
180 *Water Waste Treatm.* **10** (July 1964) 77.
181 STREATFIELD, E. L., *Chem. Ind.* **33** (1966) 569.
182 SARD, B. A., *Chem. Ind.* **29** (1962) 1310.
183 JENKINS, S. H., *Chem. Ind.* **29** (1962) 1312.
184 Pollution by Synthetic Detergents, Reports of a Symposium, *Surveyor,* 24th June, 1967, 33–38.
185 COMMITTEE ON AIR POLLUTION, *Waste, Management and Control* (National Academy of Sciences, Washington, April 1966).
186 MORRIS, S. S., *Surveyor* (29th August, 1964) 23.
187 KONONEVA, V. A. and V. B. AKSENOVA, *Gig. Sanit.* **28** (7th July, 1963).
188 ANON., *Sulphur* (June 1964) 28.
189 ANON., *Chem. Engng* **71** (1964) 92.
190 ANON., *Sulphur* (June 1964) 31.

191 AUSTIN, P. R. and S. W. TIMMERMAN, *Design and Operation of Clean Rooms* (Business Publ. Co., Detroit, 1965).
192 KUTNEWSKI, F., *J. Amer. Assoc. Contam. Control* **3** (Aug. 1965) 11.
193 FEDERAL STANDARD No. 209, *Clean Room and Work Stations Requirements, Controlled Environment* (US Govt Gen. Serv. Admin., Business Service Centre, Washington).
194 *New Scientist* **32** (13th Oct., 1966) 23.
195 LEBEDEVA, G. N. *et al., Coke Chem. USSR* **3** (1966) 32–37.
196 FISCHER, F. and L. KLOCKNER, *Wasser, Luft Betrieb* (1966) 833–835.
197 GHERSIN, Z., *Rev. chim., Bucharest* **18** (1967) 112–113.
198 KÖRBITZ, H. G. and W. BROCKE, *Schriftenreihe der Landesanstal für Immissionsschutz, Nordrhein-Westfalen* No. 10 (1968) 16–26.
199 FEUERSTEIN, D. L. *et al., Tappi* **50** (1967) 258–262.
200 BRINCK, D. L. *et al., Tappi* **50** (1967) 276–285.
201 BETHGE, P. O. and L. EHRENBERG, *Svensk Papperstidn.* **70** (1967) 347–350.
202 PRETORIUS, S. T. and W. G. MANDERSLOOT, *Powder Technol.* **1** (1967) 129–133.
203 SUZUKI, T., *Japan Air Cleaning Assoc., Tokyo* **2** (Feb. 1964) 1–4.
204 DEMIDOWICZ, A. *et al.,* Polish Patent 49 789 (31st Aug., 1965).
205 CROWE, M. J., *JAPCA* **18** (1968) 154–157.
206 SULLIVAN, D. C. *et al., Atmos. Envir.* **2** (1968) 121–133.
207 HEMEON, W. C. L., *JAPCA* **18** (1968) 166–170.
208 EHHALT, D. H., *JAPCA* **17** (1967) 518–519.
209 HINO, M., *Atmos. Envir.* **2** (1968) 149–165.
210 McCRONE, W. C. *et al., The Particle Atlas* (Ann Arbor Science Publishers Inc., 1967).
211 DELLY, J. G., *Particle Analyst* **1** (1968) 10–13.
212 WATANABE, H. and T. NAKADAI, *JAPCA* **16** (1966) 614–617.
213 KOPPLIN, J. O. *et al., J. Amer. Pharm. Assoc., Sci. Ed.,* **48** (1959) 427.
214 KOPPLIN, J. O. *et al., J. Amer. Pharm. Assoc., Sci. Ed.,* **48** (1959) 521.
215 SEIYAMA, T. and S. KAGAWA, *Anal. Chem.* **38** (1966) 1069.
216 BATTELLE MEMORIAL INSTITUTE, *Envir. Sci. Technol.* **2** (1968) 89.
217 RANDERSON, D. J., *JAPCA* **18** (1968) 249–253.
218 BOER, W., *Proc. Internat. Clean Air Congress, Lond.* (1966) 79–81.
219 LOWRY, W. P. and R. W. BOUBEL, *Meteorological Concepts in Air Sanitation* (Corvallis, Oregon, 1967).
220 EFFENBERGER, E., *Präventivmed.* **11** (1966) 601–621.
221 US NAT. CENTRE AIR POLL. CONTR., *Chem. Week* **102** (24th Feb., 1968) 32.
222 KLEIN, H., *Energie Technik* **18** (1966) 228–235.
223 KLEIN, H., *Chemie-Anlagen Verfahren* (1967) No. 4, 35–38; (1968) No. 1, 37–40.
224 BAKER, R. A., *J. Water Poll. Contr. Feder.* **35** (1963) 728–741.
225 AMER. SOC. TESTING MATERIALS (ASTM), *Manual on Industrial Water and Industrial Waste Water,* 2nd ed. (Philadelphia, Pa., 1959).
226 ALLEN, J. A., *Scientific Innovation and Industrial Prosperity* (Elsevier Publishing Co., Amsterdam, 1967).
227 SUMMER, W., *Ultra-violet and Infra-red Engineering,* p. 184 (Pitman, London, 1962).
228 GRAHAM, J., *Inst. Food Sci. Technol. (UK)* **1** (1968) 5–9.
229 DRISCOLL, B. J., *Inst. Food Sci. Technol. (UK)* **1** (1968) 11–15.
230 CARROLL, M. F., *Biochem. J.* (July, 1964).

231 MAHLER, E. A., *Proc. Internat. Clean Air Congr., Lond.* (1966).
232 MENCHER, S. K., *Petrochem. Engng* **39** (May 1967) 21-24.
233 WORLD HEALTH ORGANIZATION (WHO) *Air Pollution* (Geneva, 1961).
234 CHUTE, A. E., *Petrochem. Engng* **39** (June 1967) 32-36.
235 TERMEULEN, M. A., *Proc. Internat. Clean Air Congr., Lond.* (1966) 92-95.
236 PORTER, R. A., *Proc. 7th World Petroleum Congr., Mexico* (1967) **9**.
237 US ENVIR. HEALTH AND SAFETY RES. ASSOC., *Envir. Sci. Technol.* **1** (1967) 968.
238 FIALA E. and E. G. ZESCHMANN, *Z.V.D.I.* **109** (1967) 1139-1141.
239 EISENBUD, M. and H. G. PETROW, *Science* **144** (1964) 288.
240 CANNON, L. H. and J. M. BOWLES, *Science* **137** (1962) 765.
241 KEHOE, R. A., *Ethyl News* 1962 (Lead Industries Assoc., New York).
242 SCORER, R., *Air Pollution* (Pergamon Press, London, 1968).
243 IMAI, M. *et al., Japan J. Hyg., Tokyo* **22** (1967) 323-335.
244 ENVIRONMENTAL SCIENCE SERVICES CORP., *Solvent Emission Control Laws and the Coatings and Solvents Industry* (Stamford, Conn., 1967).
245 REINLUFT, *Chem. Engng* **74** (1967) 94-98.
246 POTTER, A. G. *et al.,* Paper to 154th Meeting A.C.S. (Div. Petroleum Chem.), Chicago, Sept. 1967.
247 NASR, A. N. M., *J. Occup. Med.* **9** (1967) 589-597.
248 SCHEEL, L. D., *Toxicology of Carbon Disulphide, Proc. Symp., Prague, 1966*, 107-115 (publ. 1967).
249 TOYAMA, T. and H. SAKURAI, *Toxicology of Carbon Disulphide, Proc. Symp., Prague, 1966*, 197-204 (publ. 1967).
250 DJURIC, D. *et al., Toxicology of Carbon Disulphide, Proc. Symp., Prague, 1966*, 118-120 (publ. 1967).
251 SCHEEL, L. D., W. C. LANE and W. E. COLEMAN, *Amer. Industr. Hyg. Assoc. J.* **29** (1968) 93.
252 ANON., *Amer. Industr. Hyg. Assoc. J.* **29** (1968) 93.
253 ANON., *Chem. Engng* (8th April, 1968) 78-79.
254 MARSH, K. J. (Stichting Concawe, The Hague, Jan. 1968).
255 YAMASHITA, K. *et al., Bunseki Kagaku* **16** (1967) 1251-1253.
256 BLOKKER, P. C., *Literature Survey of Lead Emissions from Gasoline Engines* (Stichting Concawe, The Hague, 1967).
257 PARKER, C. H., *SPE Journal* **12** (Dec. 1967) 26-30.
258 FIRST, M. W., *Proc. Internat. Clean Air Congr., Lond.* (1966) Pt. 1, 188-191.
259 HORAI, Z. *et al., J. Nara Med. Assoc., Japan* **18** (1967) 1-5.
260 MOLOKHIA, M. M. and H. SMITH, *Arch. Envir. Health* **15** (1967) 745-750.
261 HORAI, Z. *et al., J. Nara Med. Assoc., Japan* **18** (1967) 5-6.
262 POLYSCIENCE CORP., *Envir. Sci. Technol.* **2** (March 1968) 13.
263 DATEO, G. P. *et al., Food Research* **22** (1957) 440.
264 ANDERSON, W. T., *J. Opt. Soc. Amer.* **32** (1942) 121.
265 *Engineer's Digest* (Feb. 1954) 42.
266 COHEN, A. J., and H. L. SMITH, *Science* **137** (1962) 981.
267 SWIFT, D. L., *Nature* **218** (1968) 506.
268 SUMMER, W., *Ultra-violet and Infra-red Engineering* (Pitman, London, 1962).
269 SUMMER, W., *Physical Laboratory Handbook Angerer-Ebert* (Pitman, London, 1966).
270 EMMETT, P. H., *Catalysis,* 7 vols (Reinhold, New York, 1960).
271 TAYLOER, E. F. and F. T. BODURTHA, *Industrial Wastes* (August 1960).

272 MOLOS, J. E. *JAPCA* **11** (Jan. 1961) 9.
273 GOODELL, P. H., *JAPCA* **10** (1960) 234.
274 FIFE, J. A., *Heating, Piping, Air Condit.* **38** (Nov. 1966) 93–100.
275 MCATEER, D. J., *Proc. Biochem.* **3** (Apr. 1968) 60–62.
276 GANE, R. *et al.*, DSIR: *Food Investigation Technical Papers* No. 3 (HMSO, London, 1953).
277 HALL, E. G., *Food Preservation Quarterly* **15** (Apr. 1955) 66.
278 HOWARD, A., *Food Preservation Quarterly* **20** (1960) 1.
279 BRIEGER, H., *Toxicology of Caron Disulphide, Proc. Symp., Prague 1966*, 27–31 (in English).
280 BRAVO, A. H. and J. P. LODGE, *Proc. Congr. Mundial Contam. Aire, Buenos Aires*, **1** (1965) 91–99.
281 HUMBLE OIL AND REFINING CO., *Instrum. Technol.* **15** (Apr. 1968) 28–30.
282 BRUMMAGE, K. G., *Atmos. Envir.* **2** (1968) 197–224.
283 MIN. OF TECHNOL. (formerly DSIR), *Methods for the Detection of Toxic Gases in Industry* (HMSO, London, 1966).
284 SUMMER, W., *Photosensitors* (Chapman & Hall, London, 1957).
285 BURHOUSE, W. A., *Air Engng* **10** (March 1968) 18–22.
286 KOZEL, J. and V. MALY, *Staub* **28** (1968) 246–248.
287 CROCKFR, E. G. and L. F. HENDERSON, *Amer. Perfumer* **22** (1927) 325, 356.
288 ILLINOIS INST. TECHNOL., *Chem. Engng* **75** (11th March, 1968) 82.
289 OKRESS, E. C., *Trans. Microwave: Theory and Techniques* **13** (1965) 703.
290 UNGERER, W. G. and R. B. STODDARD, *Chem. Abstr.* **16** (1922) 2384.
291 TEUDT, H., *Chem. Abstr.* **14** (1920) 1685.
292 TEUDT, H., *Physiol. Abstr.* **4** (1920) 484.
293 BECK, L. H. and W. R. MILES, *Science* **106** (1947) 511.
294 BECK, L. H. and W. R. MILES, *Proc. Nat. Acad. Sci. US* **35** (1949) 292.
295 ALLISON, A. R. *et al.*, US Patent 2 849 291 (1958).
296 ALUMINIUM DEVELOPMENT ASSOCIATION, London, *Inform. Bull.* **21** (1955).
297 ALKSEEVA, M. V. and V. A. KHRUSTALEVA, *Gig. Sanit.* **25** (May 1960) 10.
298 ALLEN, F. and M. SCHWARTZ, *J. Gen. Physiol.* **24** (1940) 105.
299 GLASSTONE, S., *Physical Chemistry* (Macmillan, London; van Nostrand, New York).
300 POWELL, R. R., *Ice and Refrig.* **88** (1935) 99.
301 DUANE, J. P. and J. E. TYLER, *Interchem. Rev.* **9** (1950) 25.
302 GRANT, G. R. M., *Proc. Roy. Soc. Queensland* **60**, (1948) 93.
303 ELSBERG, *Bull. Neurol. Inst. New York* **4** (1936) 544–545.
304 KATZ, S. H. and E. J. TALBERT, *US Bureau of Mines, Techn. Paper* No. 480 (1930).
305 *Determination of Odour Thresholds for Organic Compounds* (Arthur D. Little, Inc., 1968).
306 KANO, R., *Japan Air Cleaning Assoc., Tokyo*, **2** (1964) 54–59.
307 WANNER, H. and A. GILGEN, *Arch. hyg. Bakteriol., Munich* **150** (1966) 78–91.
308 WENT, F. W., *Desert Res. Inst., Reprint Ser.* **31** (1966).
309 WENT, F. W., *JAPCA* **17** (1967) 579–580.
310 SMITH, G. T., *Rocky Mountain Med. J.* **64** (March 1967) 55–58.
311 GOETZ, A. and O. J. KLEJNOT, *JAPCA* **17** (1967) 602–603.
312 EHHALT, D. H., *JAPCA* **17** (1967) 518–519.

313 WORLD HEALTH ORGANIZATION (WHO), *International Standards for Drinking Water* (Geneva, 1963), p. 29.
314 BROOKS, W. B., *Material Prot.* **7** (Feb. 1968) 24–26.
315 JEDRZEJOWSKI, J., *Gaz. Woda. Tech. Sanit.* **33** (1959) 240 (in Polish).
316 ANON., *Chem. Engng, New York* **75** (20th May, 1968) 70.
317 SCHWANECKE, R., *Wasser Luft Betrieb* **12** (1968) 221–223.
318 ANON., *Chem. Engng, New York* **74** (6th Nov., 1967) 124–126.
319 ANON., *Chem. Engng, New York* **75** (22nd Apr., 1968) 68–69.
320 SUMMER, W., *J. Inst. Heat. Ventil. Engrs* **34** (1966) 229–236.
321 SUMMER, W., *Process Biochem.* **3** (1968) 53–56.
322 SUMMER, W., *Heizung & Lüftung, Zürich* **35** (1968) 26–28.
323 BAUM, W. A., *Bull. Amer. Meteorol. Soc.* **49** (1968) 234–237.
324 *Report: Bull. Amer. Meteorol. Soc.* **49** (1968) 237–241.
325 HUEY, NORMAN A. *et al., JAPCA* **13** (April 1963).
326 JAPIKSE, B., *Ann. New York Acad. Sci.* **116** (1964).
327 LARKINS, S. C. and W. L. DAVIS, *Industry Power* (May 1957).
328 LINDSAY, C. W. and C. NELSON, *Food Processing* (August 1954).
329 *Food Engineering* (March 1959) 60.
330 MEINHOLD, TED F. and D. D. WALKER, *Chemical Processing* (July 1956).
331 CLARK, GEORGE H. and TED F. MEINHOLD, *Chem. Processing* (June 1954).
332 BARNEBEY, H. L., DON LEE and W. J. MILLER, *ASHRAE Summer Meeting,* 1962.
333 BARNEBEY, H. L. and W. L. DAVIS, *JAPCA* (Aug. 1957).
334 BARNEBEY, H. L. and W. L. DAVIS, *Chem. Engng* (29th Dec., 1958).
335 JAPIKSE, B., *September 1967 Meeting of the East Central Section of the APCA.*
336 BARNEBEY, H. L., *Heating, Piping, Air Condit.* (1958) 153–160.
337 FISCHER, F. P., D. VERMEULEN and J. G. EYMERS, *Arch. Augenheilkunde* **109** (1935) 462–467.
338 WEAVER, E. R., *Chem. Age* (1953) 443.
339 HAINER, R. M., *Amer. Chem. Soc. 125th Nat. Meeting, Kansas City, 1954.*
340 DEPT. OF SCI. & IND. RES., *Report of the Chem. Laboratory,* The Prevention of Tarnish on Silver and Copper and their Alloys, HMSO, London, 1960.

Appendix

ORGANISATIONS

(A) Centres of Information on Air Pollution

PENNSYLVANIA STATE UNIVERSITY, CENTRE FOR AIR ENVIRONMENT STUDIES, 226 Chemical Engineering II, University Partk, Pennsylvania 16802.

A list of current work is issued annually in the 'Index to Air Pollution Research', also giving mailing addresses where additional information about a project may be obtained.

CENTRE INTERPROFESSIONNEL TECHNIQUE D'ETUDES DE LA POLLUTION DE L'AIR, 28, rue de la Source, Paris, 16e.

Review of books on the subject of air pollution CA/DOC 104 (Oct. 1966–Jan. 1967) 62 pp.

List of Foreign Organizations concerned with air pollution. CA/DOC 105, 29 pp.

INSTITUT FRANÇAIS DES COMBUSTIBLES ET DE L'ENERGIE.

Issues a monthly list of abstracts of papers published under the title 'Bulletin synoptique de Documentation Thermique'. Some 140 journals are covered published in the Argentine, Australia, Austria, Belgium, Canada, China, France, Germany, Great Britain, Italy, Japan, Poland, Roumania, Switzerland, USSR, USA, Yugoslavia. Section D applies to Air Pollution.

The list is available from the CITEPA above.

WORLD HEALTH ORGANIZATION (WHO), Geneva.

A special chapter of the Annual Reports deals with environmental health with a section on environmental pollution.

*(B) Some Official and Industrial Organizations
interested in Osmogenic Air Pollution*

INTERNATIONAL

WORLD HEALTH ORGANIZATION (WHO), AIR POLLUTION DIVISION. 8 Scherfigsvej, Copenhagen, Denmark.

FÉDÉRATION EUROPÉENNE DU GENIE CHIMIQUE, AIR POLLUTION DIVISION. Secretariat: The Institution of Chemical Engineers, 16 Belgrave Sq., London SW1.

ORGANISATION DE COOPÉRATION ET DE DÉVELOPPEMENT ECONOMIQUE: AIR POLLUTION COMMITTÉE. 2, rue André Pascal, Paris, 16e.

UNION INTERNATIONALE DES ASSOCIATIONS POUR LA PREVENTION DE LA POLLUTION DE L'AIR. 62, rue de Courcelles, Paris, 8e.

UNITED KINGDOM

ALBRIGHT & WILSON, MFG., LTD., 1 Knightsbridge Green, London SW1.

AGRICULTURAL RESEARCH COUNCIL, Ditton Laboratory, Larkfield, Maidstone, Kent.

LOW TEMPERATURE RESEARCH STATION (Agricultural Research Council and University of Cambridge), Downing Street, Cambridge.

BRITISH CELLOPHANE LTD., Bath Road, Bridgwater, Somerset.

BRITISH FOOD MANUFACTURING INDUSTRIES RESEARCH ASSOCIATION, Randalls Road, Leatherhead, Surrey.

BRITISH JUTE TRADE RESEARCH ASSOCIATION, Kinnoull Road, Kingsway West, Dundee.

CADBURY BROS LTD., Bournville, Birmingham.

J. E. CLAPHAM & SONS (WATFORD) LTD., Station Estate, Watford, Herts.

COATES BROS & CO. LTD., Easton Street, Roseberry Avenue, London WC1.

W. C. EVANS & CO. (ECCLES) LTD., Complex Works, Chadwick Road, Eccles, Manchester.

FIELD & CO. (AROMATICS) LTD., Stonefield Close, Ruislip, Middlesex.

HEATING AND VENTILATING RESEARCH ASSOCIATION, Old Bracknell Lane, Bracknell, Berks.

MEDICAL RESEARCH COUNCIL, Air Pollution Research Unit, Dunn Laboratories, St. Bartholomew's Hospital, London EC1.

NORTH WESTERN GAS BOARD, Technical Planning Division, Bradford Road, Manchester 10.

DEPT. OF TRADE AND INDUSTRY, Warren Spring Laboratory, Gunnels Wood Road, Stevenage, Herts.

CENTRAL ELECTRICITY GENERATING BOARD, Research and Development Department, Operations Branch, 24/30 Holborn, London EC1.

IMPERIAL COLLEGE OF SCIENCE AND TECHNOLOGY, ENVIRONMENTAL STUDIES DEPT., Exhibition Road, South Kensington, London SW7.

MINISTRY OF HEALTH, Alexander Fleming House, London SE1.

MINISTRY OF HOUSING AND LOCAL GOVERNMENT: ALKALI INSPECTORATE, Whitehall, London SW1.

NATIONAL SOCIETY FOR CLEAN AIR, Field House, Breams Buildings, London EC4.

OUTSIDE THE UK

Argentine

ASOCIACIÓN ARGENTINA CONTRA LA CONTAMINACIÓN DEL AIRE, Sarmiento 680, Buenos Aires, Argentine.

Australia

DEPARTMENT OF PUBLIC HEALTH: AIR POLLUTION CONTROL BRANCH, 86–88 Georges Street North, Sidney, New South Wales.

Austria

TECHNISCHER ÜBERWACHUNGSVEREIN, Strohgasse 21 a, Vienna III.

ÖSTERREICHISCHE GESELLSCHAFT FÜR INDUSTRIELLE MEDIZIN, Kinderspitalgasse 15, Vienna IX.

Belgium

CENTRE BELGE D'ETUDES ET DE DOCUMENTATION SUR L'AIR (CEBEDAIR), 2, rue Armand Stévart, Liège.

CONSEIL NATIONAL CONTRE LA POLLUTION DE L'AIR ET LE BRUIT, 169, rue de Flandre, Bruxelles.

Canada

DEPARTMENT OF NATIONAL HEALTH AND WELFARE: OCCUPATIONAL HEALTH DIVISION, 200, rue Kent, Ottawa.

Czechoslovakia

HYGIENE INSTITUTE, Ul CSL Armady 52, Bratislava, ČSR.

Denmark

AKADEMIET FOR DE TEKNISKE VIDENSKABER (Committee 62 'Air Pollution'), Røgndvalget 62, Rigensgade 11, Copenhagen 4.

France

CENTRE INTERPROFESSIONNEL TECHNIQUE D'ETUDES DE LA POLLUTION DE L'AIR (CITEPA), 28, rue de la Source, Paris 16e.

Germany

INSTITUT FÜR GEWERBLICHE WASSERWIRTSCHAFT UND LUFT REINHALTUNG, Habsburgerring 2–12, Köln.

INSTITUT FÜR WASSER, BODEN, UND LUFT HYGIENE, Correnz Platz 1, Berlin-Dahlem.

LANDESANSTALT FÜR IMMISSIONS- & BODENNÜTZUNGSSCHUTZ DES LANDES NORDRHEIN-WESTFALEN, 160 Eststrasse, Essen.

VEREIN DEUTSCHER INGENIEURE (VDI Kommission Reinhaltung der Luft), Prinz Georg Strasse 77–79, Düsseldorf 10.

Holland

TOEGEPAST NATUURWETENSCHAPPELIJK ONDERZOEK (TNO): AIR POLLUTION RESEARCH INSTITUTE, Apeldoornselaan 49, The Hague.

VOLKSGEZONDHEID VOOR DE HYGIENE VAN HET MILIEU, Nassau Dillenburgstraat 5, The Hague.

Italy

UNITED NATIONS: CONTROL OF NOISE AND FUMES (NANS), via Garibaldi 25, Turin.

Japan

JAPAN AIR CLEANING ASSOCIATION, Noson Kogyo Kaikan Bldg. No. 4, 1-Chome, Kanda-Jinbocho, Chiyodaku, Tokyo.

CONTROL OF PUBLIC NUISANCES, Tokyo-to Shutoseibi-Kyoku-Toshiko-gaibu, Kankyoka, 1-3 Chome, Narunouchi, Chiyodaku, Tokyo.

Malta

NATIONAL SOCIETY FOR CLEAN AIR, Malta.

Poland

STATE INSTITUTE OF HYGIENE, 24 Choamska, Warsaw.

Portugal

DIRECCIÃO-GERAL DOS SERVICOS INDUSTRAIS, 4a Repartição-Seguranca Industrial, R. Jose Estevâo 83 A, Lisbon.

Sweden

STATENS LUFTVÅRDSNÄMND, Sveavägen 166, Stockholm V.

United States of North America

AIR POLLUTION CONTROL ASSOCIATION (APCA), 4400 Fifth Avenue, Pittsburgh 15213, Pa.

US DEPARTMENT OF HEALTH, EDUCATION AND WELFARE (HEW), National Centre for Air Pollution Control, Technical Information Service (APTIC), 4676 Columbia Parkway, Cincinnati 26, Ohio.

ROBERT A. TAFT SANITARY ENGINEERING CENTRE, 4676 Columbia Parkway, Cincinnati 26, Ohio.

LOS ANGELES DISTRICT: AIR POLLUTION CONTROL, 434 South San Pedro Street, Los Angeles 13, Calif.

CITY OF CHICAGO: DEPARTMENT OF AIR POLLUTION CONTROL, 320 W. Clark Street, Chicago 10, Ill.

NEW YORK STATE: AIR POLLUTION CONTROL BOARD, 84 Holland Avenue, Albany 8, New York.

OREGON STATE AIR POLLUTION AUTHORITY, 1400 SW 5th Avenue, Portland Oregon.

PENNSYLVANIA STATE UNIVERSITY: CENTER FOR ENVIRONMENTAL STUDIES, 1600 Woodlands Road, Abington, Philadelphia, Penn.

Index

Abattoir waste 105
Acetaldehyde 127
Acetone 156
Acetylmethyl carbinol 66, 126
Achromobacter 132
Acrolein 67, 111, 126, 152
Activated carbon 163, 172, 183,
        184, 185
Aerobic fermentation 159
Aerosols 31, 40, 162
    deposition in lungs 41
Air, composition of 62
    exhaustion date 227
    filters, pressure on 180
    flow 212, 228
    pollution, a chemical analysis 1
Air Pollution Control Boards
        (USA) 80, 81
    control systems 283
    definitional model 95
    epidemiology 42
    from car exhausts 1
    from furnaces 1
Air standards 55
Aitken nuclei 61
Alcohol ethoxylates 150
Aldehydes 34
    detection 55
Algae 169
Aliphatic amines 107
    hydrocarbons 169

Alkali Act, 1906 152
    Works Order, 1966 152
Alkanes 150
Alkaptonuria 65
Alkenes 152
Alkylation of benzene 152
Alkylbenzene sulphonates 150
Alkyl radical 35
Allantoin 134
Allyl aldehydes 67
Allylamine 36
Allylpropyl disulphide 121
Alveoli 38
Amines 67
Ammonium, quaternary 203
Amoore's molecular shapes 11
Anaerobic fermentation 144
Animal feeding stuffs 105
    tankage 105, 117
Anomalous odour perception 100
Anosmia 87, 88
Anthraquinone disulphonic acid 201
Antibiotics 132, 151
Anti-corrosive coatings 260
Antiknock compound 73
Anti-pollution legislation 82
Aqueous solubility of osmogenic
        substances 167
Asparagus smell 65

Bacillariophyceae 169

Bacillus coli 148
Bacteria and smell 45, 171
  of sea water 148
  in water 62
Bacterial pigments 132
Bactericidal action of rays 133
  radiation 245
B-attenuation 49
Beaufort scale of wind velocities 224
Bee's honey, infra-red spectrum 15, 16
Benzidine test paper for ozone 50
Benzine 111
Berl saddles 171
Biochemical oxygen demand (BOD) 110, 126, 165, 166
  and flow velocity 147
  and temperature 146
  of detergents 150
Biodegradability 150
Black liquor from kraft paper process 156
Blood boilers 104
Blowfly 134
Bluebottle 134
Blueflies 105
Blue haze over mountains 61
Blue smoke 126, 127, 160
Body odour 57
Boiling point and volatility 32
Bone china 105
  calcined 105
  degreasing of 105
  phosphate 105
Brain 30
Bronchi 38
Brytalized aluminium 249
Bulbus olfactorius 30
Butanone 156
Butyl alcohol 17

Caffeine 67
Camphene 156
Camphor 11
Caprolactam 269

Captive balloons for air sampling 47
Carbinol 34
Carbon disulphide 45
  monoxide 34
Carbonyl 34
Carbonylfluoride 46
Carboxyl 34
3-carene 156
Carica papaya 126
Catalytic combustion 240
Catechol tannins 109
Catering industry 126
Chayen process 105, 112
Cheese odour 66
Chelate 201
Chemical aspects of odour 34
  cleaning 199
  scrubbing 168
Chimney design 206, 219, 224
Chlorella 169
Chlorination 170, 172
  of trade effluents 108
Chlorine dioxide 108
  in scrubbers 108
  peroxide 203
Chlorofluoromethanes 236
Chlorophyceae 169
Chlorophyll 237
Chloryl radical 236
Chromaesthesia 100
Chromatography 47
Chromogenic bacteria 132
Chromoprotein 130
Chrysalis 134
Cider, flavour of 17
Ciliary action 38
Claus process 85, 154
  for sulphur 85, 154
Clean air, specification of 59
Clean Air Act (1956) 76
Clean Room, specification of 59, 90
Cloth filter 178
Coexcitation 101
Coffee beans 127

Coke breeze 172
Cold cathode tubes for ultra-violet
    irradiation 245
Colloidal state 162
Combustion 154, 239, 241
Comfort engineering 70, 84
    threshold 29
Concentration of odour 19
Condensation nuclei 50, 61
Conjunctivitis 255
Contact materials for scrubbing 171
Cooking smells 37, 66, 121
Cooling towers 169
    water 233
Counter-current extraction 164
Cowls 122, 123, 182
Cranial nerves 28, 35
Cross-tainting 133
Cupola collector 176
Cyclohexane 267
Cyclohexen 3-one
Cyclonic spray tower 172
Cycloning 163
Cystine 108

Dairy waste 126
Dehumidifiers 191
Desalination 149
Detergents 150
Dew point 234
Dextran molecular filter 183
Diacetyl 127
Diallyl disulphide 121
Diatoms 169
Dichlorodifluoromethane 236
Diesel engine exhaust 40
Diffusion from a chimney 200
Diffusivity, definition of 169
Diisooctyl phthalate 160
Dimethyl disulphide 66, 121
    sulphide 66, 121
Dioctyl phthalate 160
Di-para-di-chloro-benzene 58
Dipterian flies 134
Diseases, smell caused by diabetes 65

Diureide of glyoxylic acid 134
Dixon gauze packing 171
Domestic refuse 2, 137
Doping of quartz 245
Dosage-area coefficient 48
Dose level 48
Drinking water standard,
    international 62
Droplets 31, 168
Dry air 179
    blood 107
    distillation 240
Dry-scrubber 118, 163, 178
Durene 11
Dustfall, permissible amount of 155
Dust filter 178
Dysphoric principle 100

Effluents, toxic 108
'Electric nose' 31, 55, 56, 77
Electric power consumption and
    industrial activity 5
Electrolytically brightened
    aluminium 249
Electronmicroscopy 47
Electrostatic precipitation of
    particulates 230
Emission Inventory of Chicago
    (1965) 127
Enamelling industry 158
Endopterygotous insects 134
Enzymes 131
Erythrocytes 130
Ethylbenzene 152
Ethylene 129
Ethylene-ozone reaction 127
Ethylmethylketone 156
Eupathetic threshold 29
Exhalation from lungs 40
Expanded metal filter 182
Extrasensorial osmic perception 101
Eyeshields for UV radiation 258

Faecal smells 66
Fasciculus olfactorius 30

Fat melters 104
  under ultra-violet irradiation 130
Fatty acids 130
Fermentation cells in waste
    conversion 139
Ferric hydroxide as catalyst 200
  sulphide 201
Fertilizer meal 107
Filter types 178–179, 182, 259
Fish friers 128, 182
  meal production 117
  processing 104
  smells 67
  tankage 117
Flammable concentrations in air
    243
Flare system 154
Flavobacterium 132
Flavour of cider 17
Floating roof tank 153
Flocculation in wet scrubbing 166
Fluorescence in quartz tubes 246
  of putrefaction 117
Flyblown waste 105, 108
Foam equipment for dust
    separation 177
Foam-scrubbing 163
Fog-filter scrubber 172
Food smells 37, 66
Freon 91, 236
Frigen 236
Fruit in cargoes and warehouses
    133
  motivation of 129
  ozone concentrations for 133
Fuller's earth 191
Fundamental odours 92
Furan 127
Furfural 127
Furfuraldehyde 127
Furfuryl alcohol 127
  mercaptan 127

Garbage tankage 117
Gas detection apparatus 49

Gaseous emissions, chemical
    reactions of 153
Gelatin 105, 108
Glue makers 104
Glycerol 152
Graphite in lungs 39
Grease filters 111, 182
  traps 107
Greases from animal sources 105
Greases as source of air pollution
    107
Ground-level concentration,
    formula for critical 209
Gut cleaners 113

Haemoglobin 130
Haines process 202
Hardening of UV tubes 247
Hastelloy 169
Heat-absorption of polluted air 71
Heat, and olfaction 88
  body as source of 68
  loss 69, 70
Helical flow technique in dry
    scrubbing 195
Helicopters as laboratories 47
Hexachloroethane 11
Hitachi process for sulphuric acid
    264
Hood, see Cowl
Hopcalite 49
Hydrogen peroxide 53
  phosphide 203
  sulphide 17, 53
  sulphide from sewage 144
Hydrosols 162
Hydroxyl 34
Hyperidrosis 65
Hypothesis of smell 10

Impact collector 174
  scrubber 173
Inconel for cooling towers 169
Indole 17, 66, 132, 143
Industrial solvents 158, 159

Infra-red osmogenic spectrum 13, 15, 48
    reflecting paint 153
Ingolstadt refinery 154
Inhalation 40
Insects, holometabolous 134
Intalox saddles 171
Intensity of smell 30
Interatomic bonds 16
International Clear Air Congress (1966) 273
International legislation 8, 84
Intestinal gas contents 66
Inversion, atmospheric 214, 275
Iodine, radioactive, as tracer 41
Ion exchange resin 183
Ionic character of chemical bonds 16
Iron catalysts 40
Irradiation chambers 251
Isopleths 48

Jasmin 17

Keratine 108
Ketene 45
Ketones 131
Kieselguhr 191
Kinpactor 173
Klebsiella pneumoniae 45
Kraft paper 156

Lacquering industry 158
Lagoons 147
Lapse rate 214, 215
Laser light in atmospheric mapping 52
Law of Nuisance 76
Lead-acetate paper 145
Lead, detection of 73
Leather processing 109
Legal aspects of air pollution 76, 97
    of smell 8, 43, 97
    of water pollution 82
Legislation, global 8, 84

Lemon pulp extraction 193
Lessing rings 171
Lidar for atmospheric mapping 52
Lime and sulphuric acid mixture 109
Lipase 131
Local Authority powers 78
Location of a smell 36, 226, 284
Los Angeles problem 215
Lucilia sericata 134
Lymph glands 38
Lymphocytic infiltration 45

Macroreticular resin 183
Macrosmatic animals 13
MAC-value 18, 43
Maggot infestation of skins 110
Maggots, breeding of 104, 134
    for medical research 136
Magnetron 11, 15
Manholes, smell from 144
Masking of smells 235
Maximum acceptable concentration (MAC) 18, 43
Meat odours 66
    processing 104
Medical aspects of air pollution 42
Membrane, mucous 17
Mercaptans 108
Mercury-vapour 49
Mercury-vapour discharge tubes 244
Merosmia 100
Metabolic odours 65, 66
Metamorphosis 134
Methaemoglobin 130
Methane 143
Methionine 152
L-S-Methyl cysteine sulphoxide 66, 121
Methyl-ethyl carbinol 127
Methyl mercaptan 152
β-Methylmercaptopropionaldehyde 66
Microbicidal radiation 245

Micrococci 132
Microspora 170
Milk flavours 67
Mineral oil 160
Mitosis inhibition by hot particles 42
Mitsubishi process for ammonium sulphate 264
Mobile laboratories 282
Molasses 160
Molecular morphology 10
  sieves 183, 184
Monel for cooling towers 169
Mould formation in fruit 130
Mouth breathing 38
Mucor Aspergillus niger 130
Multi-wash collector 175
Munsell theory of colour 17
Mustard smell 35
Myelin 101
Myxophyceae 169

Nasal air flow 36
  decongestants 88
  irritation threshold 28
  mucus 38
Nascent oxygen 244
National per capita incomes 6
Neoprene coating for cooling towers 169
Nervous pulses 44
  system, working of the 87
Neutron activation of airborne particles 49
Nitrogen oxides 54
Noise Abatement Act (1960) 76
Nucleic acid 132
Nucleoprotein 133
Nuisance in Common Law 77
  Order 78

Odour classifications 91
  concentrations, measurement of 189
  evaluation in court cases 76, 97
  in human behaviour 58
  intensity index 21
  maximum acceptable values, 18, 43
  panels 27, 37
  psychological aspects of 37
  retentivity 66
  sensation and concentration 37, 44
  social aspects of 57
Odour intensity index (OII) 21, 48, 92
Odourless factory 105, 275
Odourmania 57
Odour-pain relationship 28
Odour perception, anomalous, 100
  radiation 15
  threshold 18, 22, 23, 24, 25
Offensive trades 104
Oil refineries 152
  tanks, evaporation from 153
Oleate 111
Olfaction 100
  principle of 9
Olfactory bulb 30
  epithelium 26, 36
  hairs 12
  nerve 30
  range of man 21
  triangulation 36
Onion smell 35
Open air UV installations 258
Optical breakdown of quartz tubes 246
  filters 239
Organolepsis 244
Organoleptic assessment of smell 43
Oscillatoria 169
Osmic stimuli 100
Osmoceptors 9, 12
  different levels of stimulation 44
Osmodysphoria 100
Osmophobia 100
Osteomyelitis 134
Ostwald theory of colour 17
Oxygen depleted air in sewers 145
  requirements of river fauna 148

Ozaena 65, 88
Ozone 58, 129
  and ketene 45
  detection 50
  threshold of 45
Ozonides 61

Packed towers 171
Pain and smell 36
Palmitate 111
Panapak 171
Panel tests 86, 95
Papain 126
Papaya fruit 126
Paper mills 156
n-Paraffins 150
Parosmia 88, 100
Particle Atlas 178
Particulate matter, filtration 179
Pasteurized milk 132
Pathogenic effects of odours 38
Pease-Anthony scrubber 172
Pendant drop technique 55
Penetrability of filters 181, 182
Pentanone 156
Perchloroethylene 111
Perchloryl fluoride 236
Persistence of odours 33
Personal deodorants 65
Persorption 183, 184
Perspiration 64, 65
Petrol refineries 85
Petroleum industry 152
Phase separation 162, 178
Phenol detection 55
Phloroglucinol tannin 109
Phormia regina 134
Photism 100
Photonitrosation 267
Photosynthesis 237
Pinenes 156
Pit pegs 12
Plasticisers 160
Pneumoconiosis 39
Polychloroprene (PCP) 169

Polytetrafluoroethylene (PTFE) 46
Population trends 2
Potable effluent from sewage works
    149
  water 62
Potassium iodide ozone detection
    50
  permanganate 203
Precipitation 166, 230
Principles of olfaction 9
Projicience 30
Propanone 156
Propyl methyl ketone 156
Protein metabolism 65
Pseudomonas fluorescens 132
Psychrophilic bacteria 131
Public Health Act (1936) 78
Public Health (Drainage of Trade
    Premises) Act (1957) 107
Pulmonary deposition 40, 41
Pumpkin smell 66
Pure air 59
Pyrazolone dye 259
Pyrene 42
Pyridine 127
Pyrocatechol 160
Pyrogallol tannin 109
Pyrolysis of black liquor 156

Quartz doping 245
  envelope 244
Quinone 160

Radiation hyopthesis of odour 13
Radiochemical determinations 49
Radio-tracers 47
Raman shift 14,15
Raschig rings 171
Refrigerated cargo vessels 129
Refrigeration and bacteria 131
Refuse, domestic 137
Renal dysfunction 45
Reodorization 58
Residence time 255
Residual air in lungs 40

Respiratory air, composition of 63
Ribose nucleic acid (RNA) 133
Ringelmann chart 77
River (Prevention of Pollution) Act
    (1961) 147, 232
River Boards 147
Rooms, deodorization of 255
Ross partition rings 171
Rotary driers 118
Rylands v. Fletcher 77, 78

Salmonella 132
    in water 62
Sampling of air 47
Satellites for pollution control 272
Saturnid moth 15
Sausage skin makers 113
Scanogram 41
Scavenging of ships' holds 133
Scenedesmus 169
Scrubber condenser 118
Scrubbing towers 171
Sea water, constituents of 170
Sedimentation tanks 144, 145
Senescence and loss of oxygen 63,
    64
Senility and smell perception 87
Sensory pits 11
Septic tank 144
Serratia 132
Side-chain theory of Ehrlich 10
Silica glass 245
Silicononyl alcohol 11
Silicosis 39
Sludge drying beds 144
    gas, composition of 142, 143
Smell-blindness 11
Smog index 48
Smokeless fuel 76
Sneezing 36
Sniffing 16, 17, 30, 36
Snifter set 186
Social aspect of odour 57
    economic product 7
Sodium zeolite 183

Soil as a catalyst 406
    effect of, on water 62
    pollution 150
Solid sols 62
    waste in USA 47
Solvent recovery costs 187
Sorption 163, 183, 193
Sorptive power of activated carbon
    190
Sour odours 153
Southern Outfall Works, London
    149
Soybean extraction 189
Specific gravity of gases 229
Spectral analysis by computer 48
Spirillum 170
Sporectrichum carnis 130
Spray condenser 118
    nozzles 164
Spraypak 171
Stack height 154, 155
Stacks 206
Stale air 37, 59, 66
Steam driers 118
Stearate 111
Stefan's Law 69
Stem rot in bananas 129
Stokes-Einstein Law 169
Stretford process 201
Strontium-90 49
Styrene 152
Subjective odour sensations 37
Sulphate process for kraft paper
    156
Sutton's formula for measuring
        pollutants in plumes 207
Sweet odour 34
Synergistic processes 45

Tallow 105, 111
Tankage 117
Tanner's paint 109, 110
Tanneries 108, 109
Tannin 109
Tartrazine 259

Television in pollution control 226, 284
Tenderizers 126
Terminology of odour 17
Terpenes 60, 61
Tetrachloroethylene 111
Tetraethyl lead 73
Tetramethyl base test paper for ozone 50
Thamnidum 130
Thermal combustion 240
  oxidation 154
  power plant 85
Thermolabile sulphur compounds 121
Thin-film odour detector 56
Thiophenol mercaptan 36
Threshold of mixtures 19, 20, 21
  of perception 18
  of smell 28

Tidal air 40
Titanium dioxide 245
Tobacco smoke 31, 45
Toxicity of odours 42, 43
Trees resistant to air pollution 74
Trichlorofluoromethane 236
Trigeminal nerve 28
Trigonum olfactorium 30
Trimethylamine 17, 135, 106
Trinitroacetonitrile 11
Trinitrophenol 202
Tripe boilers 104, 115
Tungsten metasilicate 258
Turbulent wet-scrubber 175
Turpentine 157
Type designator coefficient 48

Ulothrix 169
Ultrasound as catalyst 200
Ultra-violet deodorization 244, 277
  indicator 258
  radiation 244
Urine, odour 65
  zinc in 46

Urinoid 65

Van de Graaf principle 230
Vegetation and air pollution 73
Vehicle exhausts 83
Ventilating rates 64
Ventilation as a closed circuit 125
Venturi scrubber 173
Vermicidal effect of ozone 133
Vibrio desulfuricans 170
Vibrio estuarii 170
Vibrio thermodesulfuricans 170
Vinyl alcohol 258
Viscose plant, emission from 226
  staple fibre 45
Visual rods and cones 44
Volatility 31

Wadelin's technique for ozone 50, 51
Washing tower 108
Waste control, modes of 261, 262
  conversion 137, 139
Water as coolant 276
  colour limits of 61
  pollution legislation 82
Waveguides, biological 16
Weather and air pollution 72
Weber-Fechner's law 26, 56, 90
Wet-scrubbing 163, 164
White fish 117, 118
Wien's Displacement Law 68
Wind speed, direction and height 225, 273
Wooden cooling towers 169
World Health Organisation (WHO) 61, 62

Yokkaichi asthma 265
Yttrium-90 49

Zeolite 202, 183
Zinc oxide film as odour detector 56